Atlas of Biomarkers for Alzheimer's Disease

Manuel Menéndez González

Atlas of Biomarkers for Alzheimer's Disease

 Springer

Manuel Menéndez González
Department of Neurology
Hospital Álvarez-Buylla
Mieres
Asturias
Spain

ISBN 978-3-319-07988-2   ISBN 978-3-319-07989-9   (eBook)
DOI 10.1007/978 3 319 07989 9
Springer Cham Heidelberg New York Dordrecht London

Library of Congress Control Number: 2014942059

Printed on acid-free paper

Springer is part of Springer Science+Business Media (www.springer.com)

To my grandmother Lola, who died with Alzheimer's disease. Her funny sayings still resound in my mind as the first day

Preface

Neurodegenerative diseases are strongly linked with age, with older people being at higher risk. Global population demographics show that the world is rapidly aging, and so the prevalence of neurodegenerative diseases is rising, with Alzheimer's disease (AD) leading the statistics. In view of the growing prevalence of AD worldwide, better diagnostic tools and more effective therapeutic interventions are urgently needed, and much work in this field has been done in recent decades.

A major goal of current clinical research in AD is to improve early detection of disease and presymptomatic detection of neuronal dysfunction, concurrently with the development of better tools to assess disease progression in this group of disorders. The putative correlates used for detection and assessment are commonly referred to as AD-related biomarkers. The ideal biomarker should be easy to detect and quantify, reproducible, not subject to wide variation in the general population, and unaffected by comorbid factors. To evaluate therapies, a biomarker needs to change linearly with disease progression and closely correlate with established clinicopathological parameters of the disease.

The vast number of important applications in this field, combined with the untamed diversity of already identified biomarkers, highlights the pressing need to structure the research on AD biomarkers into a solid, a comprehensive, and an easy-to-use tool to be used in clinical settings. To date, few publications have systematically compiled results on this topic, and no atlas has been published. The objective of this book is to collect and summarize the most important studies in this field. Readers can find here a guide for reviewing the current status of research while easily visualizing outstanding results. The chapters of this atlas cover early diagnosis and risk of conversion, differential diagnosis, tracking of disease progression, and application of biomarkers in clinical practice. Each of these chapters includes an introduction and sections on biomarkers in cerebrospinal fluid and neuroimaging biomarkers. Although some other biomarkers have also been explored, we focus on the most important ones.

Contents

Early Diagnosis and Risk of Conversion from Presymptomatic Stages

1

1.1 Introduction

The relevance of the early diagnosis of Alzheimer's disease (AD) relies on the hypothesis that pharmacological interventions with disease-modifying compounds are likely to produce clinically relevant benefits if started early enough in the continuum towards dementia. Currently, many potential disease-modifying therapies are being developed and evaluated at the preclinical stage and will lead to clinical trials in the near future for which biomarkers are urgently needed. Biomarkers are variables (physical, chemical, or anatomical) that can be measured in a person and reflect the state of a disease. The search for biomarkers of preclinical AD is becoming increasingly important because pathogenesis-targeted neuroprotective strategies are being developed for future use in "at risk" populations. Advances in new neuroimaging probes and technologies; identification of new biochemical markers of AD in plasma, blood, and cerebrospinal fluid (CSF); and breakthroughs in molecular genetics and basic neuroscience are gradually translating into better understanding of predisposing and pre-clinical factors that lead to progressive neurode-generation and finally to cognitive and behavioral symptoms and dementia.

The notion of preclinical AD designates cognitively normal subjects harboring early AD pathology that is not severe enough to induce significant cognitive signs and functional deterioration. There are two main predementia stages: subjective cognitive impairment (SCI) and mild cognitive impairment (MCI). SCI is defined as a person's belief that his or her thinking abilities, including memory, are not as good as they were (Fig. 1.1). It is a change people notice in themselves, without being told by others and without significant disturbance in neuropsychological assessment. It is a personal conviction that people see as a problem. The most common complaints are due to memory issues. SCI is much more common in later life than the objective problems that suggest MCI or dementia. In addition, depression is associated with subjective memory problems, as are older age, female sex, and low educational attainment. Depression is itself a risk factor for dementia, making the diagnostic task even more difficult.

SCI is not simply a characteristic of the "worried well"; it should be taken seriously. The poorer memory performance at follow-up and the association of reduced longitudinal memory performance with hypometabolism in the precuneus at baseline support the concept of SMI as the earliest manifestation of AD (Scheef et al. 2012).

MCI is the point at which others begin to recognize that something's going on with a person's memory or cognitive functions. Perhaps a doctor can detect deficits from a clinical interview, or co-workers may notice a change in performance (Figs. 1.1 and 1.2). Although MCI can present with a variety of symptoms, when memory loss is the predominant symptom it is termed *amnestic MCI* (aMCI) and is frequently

M. Menéndez González, *Atlas of Biomarkers for Alzheimer's Disease*,
DOI: 10.1007/978-3-319-07989-9_1, © Springer International Publishing Switzerland 2014

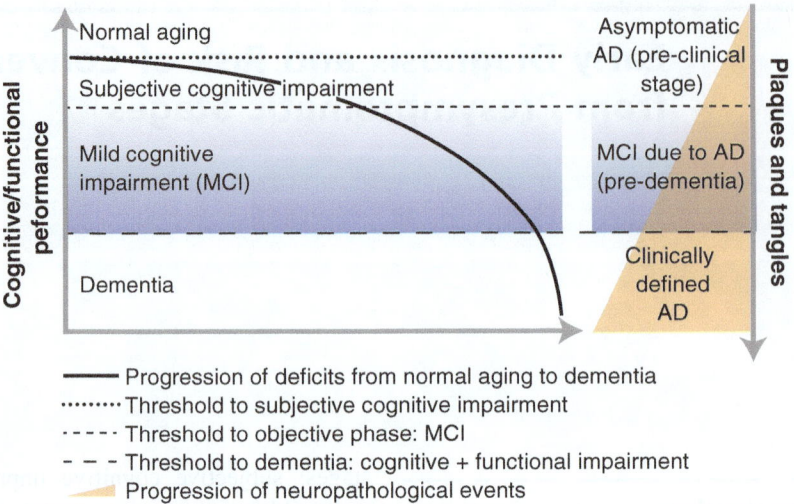

Fig. 1.1 Hypothetical model of the pathological processes in Alzheimer's disease (AD), focusing on the amyloid b peptide (Ab) cascade (from Forlenza et al. 2010)

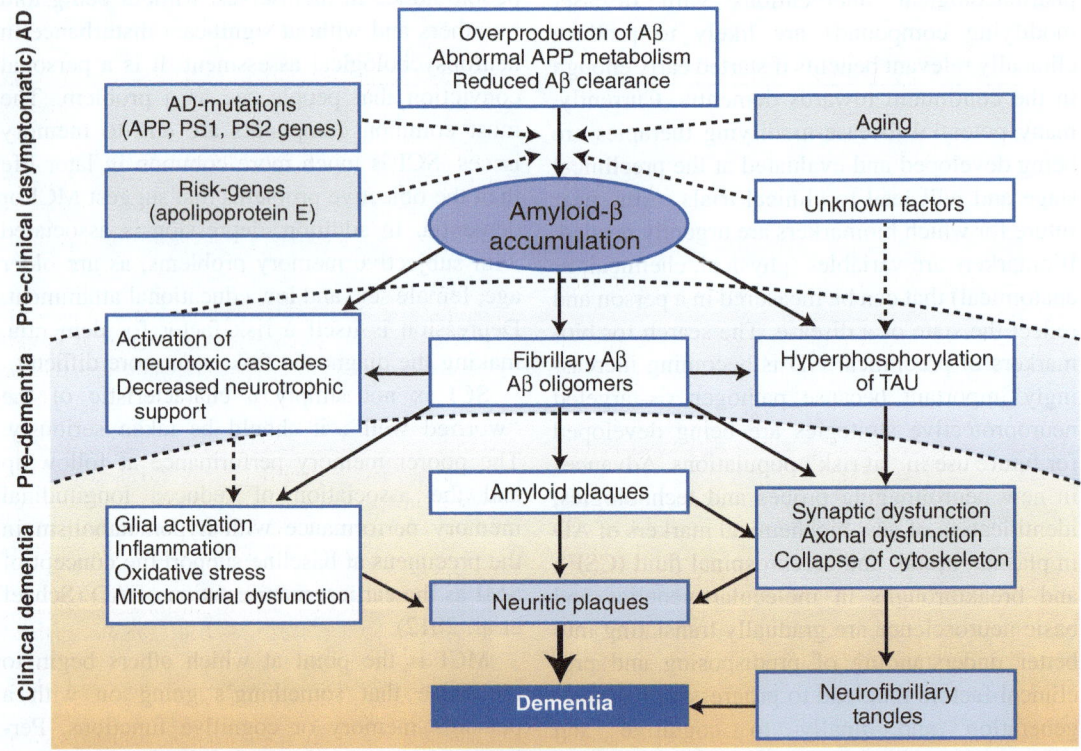

Fig. 1.2 Relationship between the progression of cognitive and functional symptoms and the neuropathological events in the transition from asymptomatic AD to mild cognitive impairment due to AD and clinically manifest dementia of the AD type. Aβ—amyloid beta; APP—amyloid precursor protein (from Forlenza et al. 2010)

Fig. 1.3 Evolutive stages from health to normal age-related memory loss (*yellow line*) or AD dementia (*red line*). The *blue line* represents the stage of mild cognitive impairment (MCI), which typically affects the memory domains while other cognitive domains are preserved. Not represented here are some normal people with age-related memory loss in whom MCI may also be diagnosed

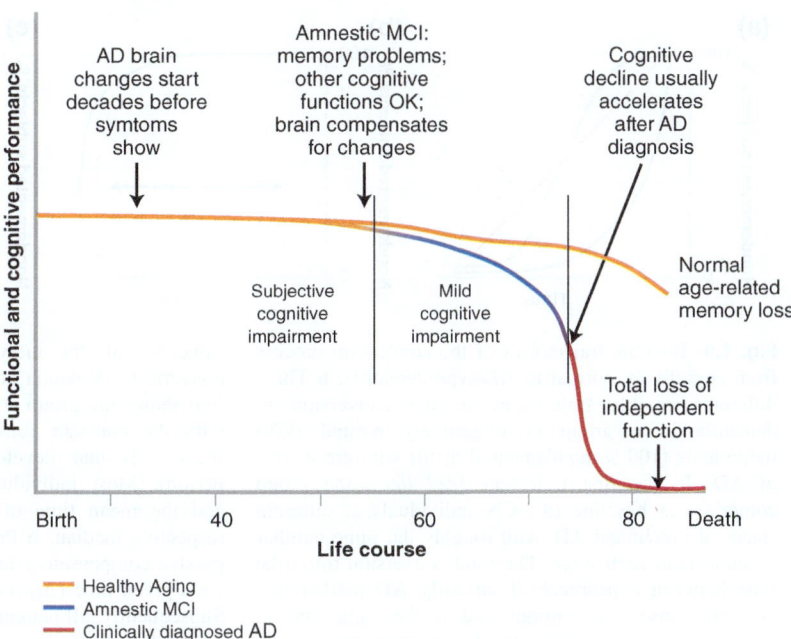

seen as a prodromal stage of AD. Cognitive deficits are found in neuropsychological assessment, although these do not have a major impact on functional capacities. Among older persons with MCI, 50–70 % also show neurobiological AD and develop dementia within the next 5–7 years. The majority of people older than 80 years develop an MCI syndrome, and half of those progress to dementia. Once the MCI syndrome is present, the symptoms of dementia appear within 2–3 years (Figs. 1.3 and 1.4). Progression from normal to MCI or from normal to MCI to dementia is not always linear; reversion from MCI is also possible and is fairly common. Those who revert remain at increased risk for future cognitive decline (Koepsell and Monsell 2012); thus individuals who develop MCI and later return to normal can subsequently progress to dementia (Lopez et al. 2012).

Diagnostic or core biomarkers express measures of the underlying molecular pathology of a disease. In AD, biomarkers are generally classified as biomarkers of amyloid accumulation and biomarkers of neurodegeneration. Core AD biomarkers reflect amyloid pathology (Aβ1-40/Aβ1-42 extracellular accumulation) or intracellular deposit of neurofibrillary tangles

(hyperphosphorylated Tau inclusions). As such, biomarkers serve to identify in living individuals a variety of neuropathological features previously detected only by the analysis of specimens from biopsy or necropsy. The availability of biomarkers for quantifying in vivo AD pathology has propelled advances in the understanding of AD as a dynamic clinicopathological entity. In contrast with the cross-sectional nature of neuropathology, biomarker assessments allow for longitudinal observations necessary to describe the temporal progression of neuropathology in neurodegenerative diseases. Indeed, the value of imaging or fluid biomarkers for supporting the diagnosis of AD in living individuals has been acknowledged in the 2011 National Institute on Aging-Alzheimer's Association (NIA-AA) criteria (*see* Chap. 4).

1.1.1 Key Concepts

1. The pathologic processes that underlie neurodegenerative diseases appear to start 10–20 years before symptoms develop. In AD, these processes can be identified using various biomarkers.

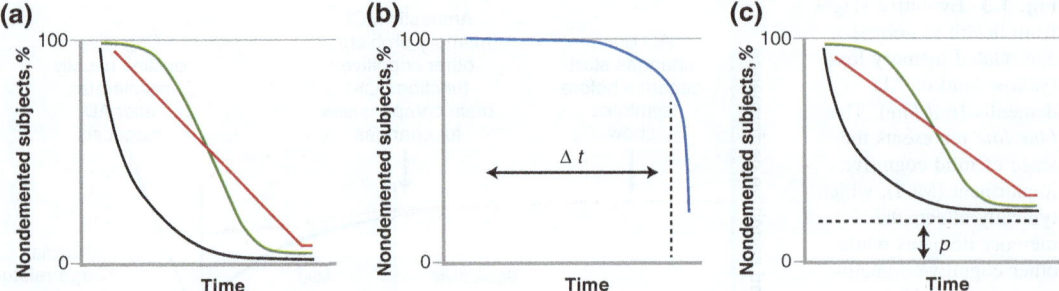

Fig. 1.4 Possible trajectories of the conversion process from cognitively normal to AD-type dementia. **a** Three different possible trajectories of the conversion to dementia in a group of cognitively normal (CN) individuals (100 % nondemented at t0) who are at risk of AD. In the first trajectory (*red line*), the group comprises at baseline (t0) CN individuals at different stages of preclinical AD, with roughly the same number of subjects at each stage. The total conversion time (the time between appearance of an early AD marker and dementia onset) is constant and is the same for all subjects (*t*), and the number of converters in a given period is constant. In the second scenario (*green line*), the group comprises people with preclinical AD, with a Gaussian distribution of the individuals at different stages of advancement (most individuals being at the intermediate stage). The total conversion time is constant and the

same for all the individuals (*t*). Most of the group converts to dementia at around t1/2. Finally, the *black line* shows the group comprising CN at preclinical AD, with the constant conversion rate (proportion of the individuals that develop dementia in a given time period). Most individuals convert to dementia early, and the mean time of conversion is higher than the respective median. **b** Preclinical AD individuals with a passive compensatory mechanism that delays conversion for a given time (*Δt*), until the mechanism is exhausted. Subsequently, all patients convert to dementia, following one of the trajectories presented on panel (**a**). **c** Preclinical AD individuals with an active compensatory mechanism that prevents conversion to dementia in a certain proportion of cases (*p*), whichever trajectory the conversion process follows (from Lazarczyk et al. 2012)

2. Biomarkers are physical, chemical, or anatomical variables that can be measured in a person and reflect the state of the disease.
3. The relevance of the early diagnosis of AD relies on the hypothesis that pharmacological interventions with disease-modifying compounds are likely to produce clinically relevant benefits if started early enough in the continuum towards dementia (predementia stages).
4. The notion of preclinical AD designates cognitively normal subjects harboring early AD pathology that is not severe enough to induce cognitive signs. There are two main predementia stages: subjective cognitive impairment (SCI) and mild cognitive impairment (MCI):
   1. SCI is defined as a person's belief that his or her thinking abilities, including memory, are not as good

as they were. It is a change people notice in themselves, without being told by others and without significant disturbance in neuropsychological assessment.
   2. MCI is the point at which others begin to recognize changes in a person's memory or cognitive functions. Deficits are found in neuropsychological assessment, although these do not have a major impact on functional capacities.

1.2   CSF Biomarkers

Examination of biological fluids provides quantitative information regarding biosynthesis, concentration, and kinetics of biomarkers (and their metabolites) that are of interest regarding

Fig. 1.5 Mean baseline and follow-up levels of CSF Tau (**a**) and CSF Aβ42 (**b**) in healthy controls (HC) who were followed for 4 years, and in two cohorts of patients with AD who were followed for 1 or 2 years. Error bars represent standard errors of the mean. This figure illustrates that differences between controls and AD patients by far surpass the within-group differences over time (from Buchhave et al. 2009)

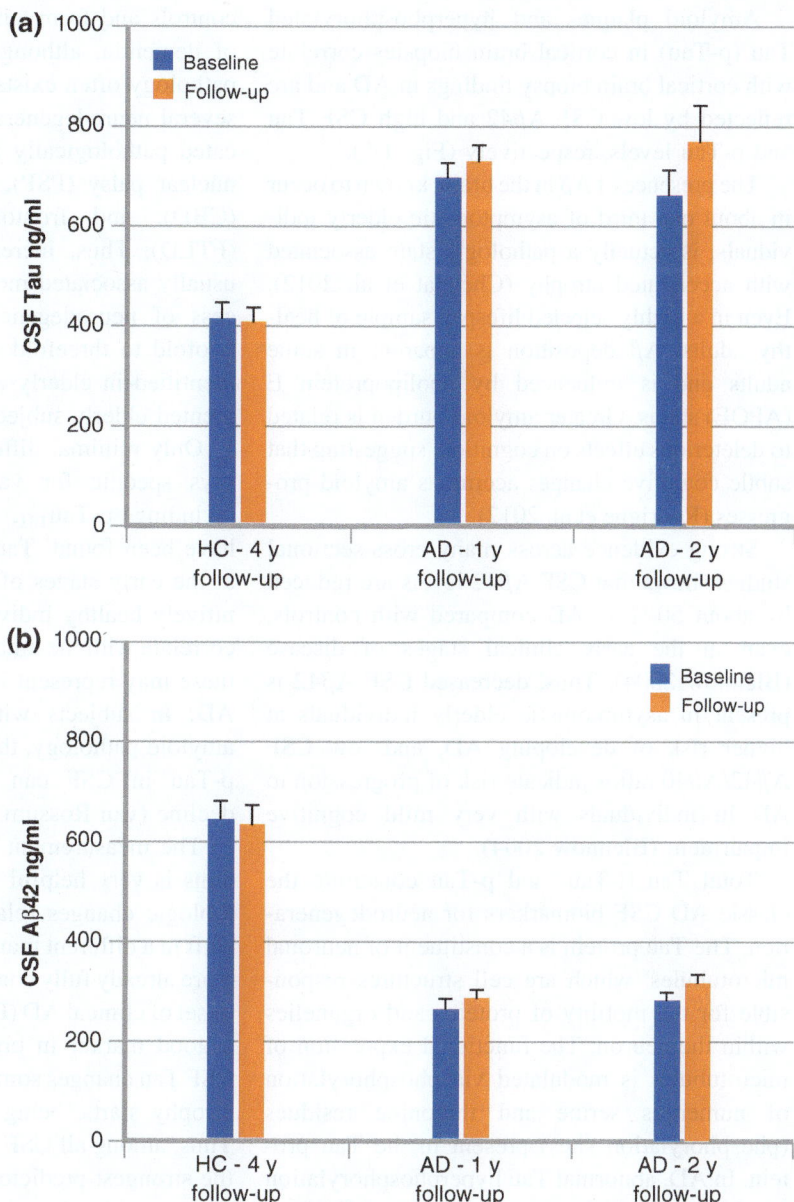

dementias. High-throughput analytical platforms are now available for detailed analysis of fluid biomarkers, and soon advanced proteomics techniques may reveal signatures for all neurodegenerative diseases. Owing to the contiguity of CSF to the brain parenchyma and its free exchange in the brain's extracellular space, the biochemical composition of CSF can provide information on brain chemistry. The distinctive features of CSF,

together with the low incidence of complications after lumbar puncture (Menéndez-González 2014a, b), have supported the introduction of lumbar puncture and analyses of CSF biomarkers into routine clinical practice in some centers.

The presence of Aβ in the brain, known to occur in about one third of asymptomatic elderly individuals, is actually a pathologic state associated with accelerated atrophy (Chételat et al. 2012).

Amyloid plaques and hyperphosphorylated Tau (p-Tau) in cortical brain biopsies correlate with cortical brain biopsy findings in AD and are reflected by low CSF Aβ42 and high CSF Tau and p-Tau levels, respectively (Fig. 1.5).

The presence of Aβ in the brain, known to occur in about one third of asymptomatic elderly individuals, is actually a pathologic state associated with accelerated atrophy (Chételat et al. 2012). Even in a highly selected lifespan sample of healthy adults, Aβ deposition is apparent in some adults and is influenced by apolipoprotein E (APOE) status. Greater amyloid burden is related to deleterious effects on cognition, suggesting that subtle cognitive changes accrue as amyloid progresses (Rodrigue et al. 2012).

Strong evidence across many cross-sectional studies shows that CSF Aβ42 levels are reduced by about 50 % in AD compared with controls, even in the early clinical stages of disease (Blennow 2004). Thus, decreased CSF Aβ42 is present in asymptomatic elderly individuals at higher risk of developing AD, and low CSF Aβ42/Aβ40 ratios indicate risk of progression to AD in individuals with very mild cognitive impairment (Blennow 2004).

Total Tau (t-Tau) and p-Tau constitute the classic AD CSF biomarkers for neurodegeneration. The Tau protein is a constituent of neuronal microtubules, which are cell structures responsible for the motility of proteins and organelles within the neuron. The functional expression of microtubules is modulated via phosphorylation of numerous serine and threonine residues (phosphorylation sites) present in the Tau protein. In AD, abnormal Tau hyperphosphorylation is observed within neurons as neurofibrillary tangles or dystrophic neurites present in neuritic plaques. In MCI, concentrations of t-Tau and p-Tau in the 181-threonine position are elevated by 30–40 %, whereas in AD patients, these elevations are as high as 40–50 %. Synaptic injury or cellular death contributes to the leakage of t-Tau and p-Tau into the extracellular space. In fact, the CSF t-Tau concentration is also increased in patients with encephalitis, trauma, and stroke. The CSF Tau concentration is useful for distinguishing AD patients from controls and from subjects with non-AD forms of dementia, although overlap at the level of pathology often exists. Tau has a central role in several neurodegenerative diseases being implicated pathologically in AD, progressive supranuclear palsy (PSP), corticobasal degeneration (CBD), and frontotemporal lobe dementia (FTLD). Thus, increased CSF Tau levels are usually associated more generally with the process of neurodegeneration. An approximately twofold to threefold increase of tTau has been identified in elderly AD patients versus nondemented elderly subjects (Blennow et al. 2001).

Only minimal differences among immunoassays specific for various epitopes of p-Tau, including p-Tau$_{181}$, p-Tau$_{231}$, and p-Tau$_{199}$, have been found. Tau elevation seems to occur at the early stages of disease and in some cognitively healthy individuals, in whom its levels correlate with the amount of amyloid deposition; these may represent individuals with preclinical AD. In subjects with MCI and evidence of amyloid pathology, the injury markers t-Tau and p-Tau in CSF can predict further cognitive decline (van Rossum et al. 2012).

The measurement of these two key AD proteins is very helpful for detecting the neuropathologic changes related to AD, although they shift in a different manner: CSF levels of amyloid β are already fully changed 5–10 years before the onset of clinical AD (Buchhave et al. 2012), being a good marker in predementia stages, whereas CSF Tau changes some time later, when the brain atrophy starts, being a good marker of injury. Thus, among all CSF biomarkers, low Aβ42 was the strongest predictor of clinical progression in patients with subjective complaints. These results are in line with the hypothesis that the cascade of pathologic events starts with deposition of Aβ42, whereas neuronal degeneration and hyperphosphorylation of Tau are more downstream events, closer to clinical manifestation of AD (van Harten et al. 2013) (Fig. 1.2).

The diagnostic accuracy of these CSF biomarkers has also been substantiated in analyses in which the diagnosis was then proven by autopsy, with discriminatory power comparable or superior to that of studies in patients with

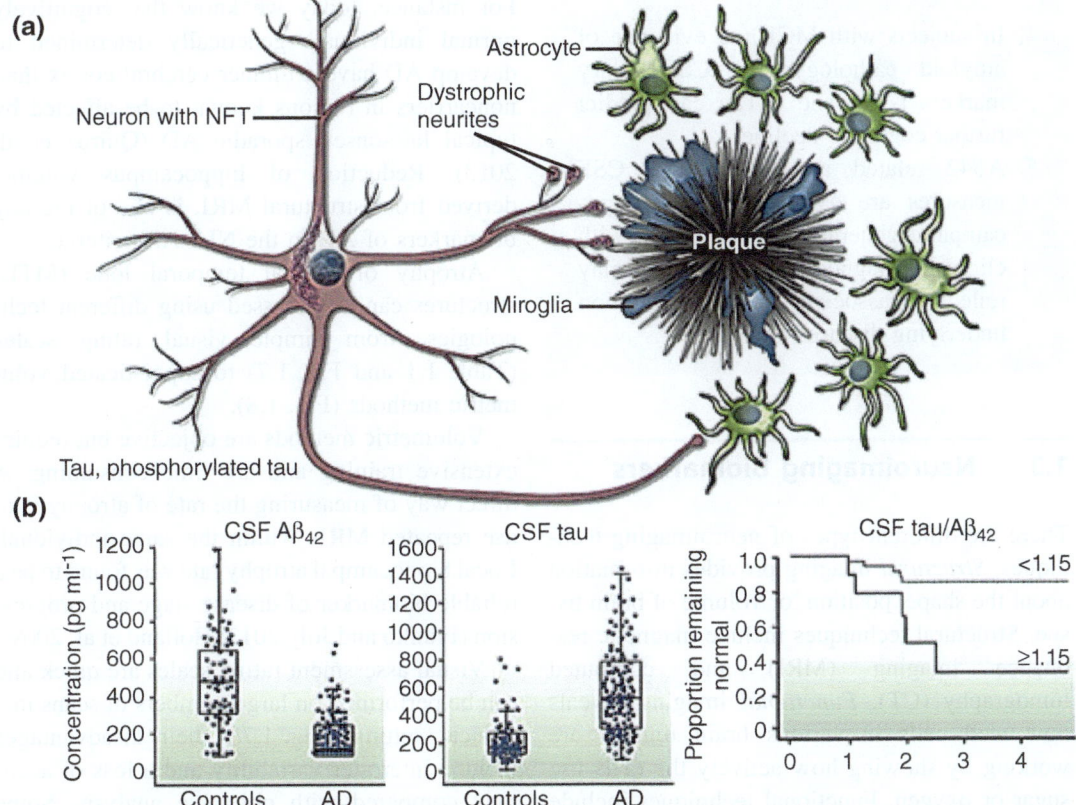

Fig. 1.6 Pathophysiology of Alzheimer's disease. Aβ42 leads to amyloid accumulation into extracellular amyloid plaques while p-Tau leads to intracellular neurofibrillary tangles (NFT). Both Aβ42 and p-Tau can be measured in CSF. Results differ between controls and patients with AD. In follow-up studies, individuals with a low Tau/Aβ42 ratio do not convert to AD over time, whereas those with a high ratio do

clinical diagnoses only. Despite favorable results obtained from large cohorts of dementia patients, however, translation of these technological advances into diagnostic methods is limited by important factors such as reliability (that is, variability across laboratories) and accuracy (that is, interindividual biological variability) (Fig. 1.6).

1.2.1 Key Concepts

1. Numerous studies already suggest that CSF biomarkers have a high potential as diagnostic tools in AD: decreased levels of β-amyloid1-42 (Aβ42) and increased total Tau protein (t-Tau) and Tau phosphorylated at position threonine 181 (p-Tau$_{181}$) in CSF have been established as valid biomarkers for the diagnosis and prognosis of AD.

2. Despite favorable results obtained from large cohorts of dementia patients, however, translation of these technological advances into diagnostic methods is limited by important factors such as reliability (variability across laboratories) and accuracy (interindividual biological variability).

3. The Tau/Aβ42 ratio predicts future cognitive decline with high accuracy at all ages.

4. In subjects with MCI and evidence of amyloid pathology, the CSF injury markers t-Tau and p-Tau can predict further cognitive decline.
5. $A\beta42$ related and Tau-related CSF measures are associated with hippocampal degeneration in individuals with clinically diagnosed early AD and may reflect an association with a common underlying disease mechanism.

1.3   Neuroimaging Biomarkers

There are different types of neuroimaging techniques. *Structural* imaging provides information about the shape, position, or volume of brain tissue. Structural techniques include magnetic resonance imaging (MRI) and computed tomography (CT). *Functional* imaging reveals how well cells in various brain regions are working by showing how actively the cells use sugar or oxygen. Functional techniques include positron emission tomography (PET) and functional MRI (fMRI). *Molecular* imaging uses highly targeted radiotracers to detect cellular or chemical changes linked to specific diseases. Molecular imaging technologies include variants of PET and single photon emission computed tomography (SPECT) with different radioligands.

1.3.1   MRI

Structural imaging is an integral part of the clinical assessment of patients with suspected AD. The sole objective of neuroimaging studies in dementia used to be ruling out large damage in the brain such as stroke, tumors, or hydrocephalus. However, prospective data on the natural history of change in structural markers from preclinical to overt stages of AD radically changed how the disease is conceptualized, and is influencing its future diagnosis and treatment.

For instance, today we know that cognitively normal individuals genetically determined to develop AD have a thinner cerebral cortex than noncarriers in regions known to be affected by typical late-onset, sporadic AD (Quiroz et al. 2013). Reduction of hippocampus volume, derived from structural MRI, is one of the key biomarkers of AD in the NIA-AA criteria.

Atrophy of medial temporal lobe (MTL) structures can be assessed using different technologies, from simple visual rating scales (Table 1.1 and Fig. 1.7) to sophisticated volumetric methods (Fig. 1.8).

Volumetric methods are objective but require extensive training and are time consuming. A direct way of measuring the rate of atrophy is to use repeated MRIs within the same individual. Local hippocampal atrophy rate was found to be a reliable biomarker of disease stage and progression (Frankó and Joly 2013; Holland et al. 2009).

Visual assessment rating scales are quick and can be performed on large numbers of scans in a clinical setting (Fig. 1.7); their disadvantages include interrater variability and a loss of accuracy compared with objective analysis. Some studies have found that visual rating assessment of the MTL gave similar prediction accuracy to multivariate classification and manual hippocampal volumes, but others reported that the visual rating assessment failed to detect patients at high risk, such as people carrying mutations of familial AD, and also failed to detect progression over time (Duara et al. 2013; Westman et al. 2011).

Relatively easy but objective two-dimensional methods such as the Medial Temporal Lobe Atrophy index (MTAi) (Fig. 1.9) have also been developed and proven to be valid. The MTAi compares the extent of atrophy in the MTL with the extent of global brain atrophy (Menéndez-González et al. 2014). Some other indices (Table 1.2) using volumetry have also been described to compare the extent of atrophy in the MTL with the extent of global brain atrophy in three dimensions (Giesel et al. 2006; Menéndez-González, 2014a, b).

Table 1.1 Medial temporal lobe atrophy visual rating scale[a]

Score	Width of choroid fissure	Width of temporal horn	Height of hippocampus
0	Normal	Normal	Normal
1	↑	Normal	Normal
2	↑↑	↑	↓
3	↑↑↑	↑↑	↓↓
4	↑↑↑	↑↑↑↑	↓↓↓

[a] A score of 0–4 is given separately for the left and right sides

↑—increase

↓—decrease

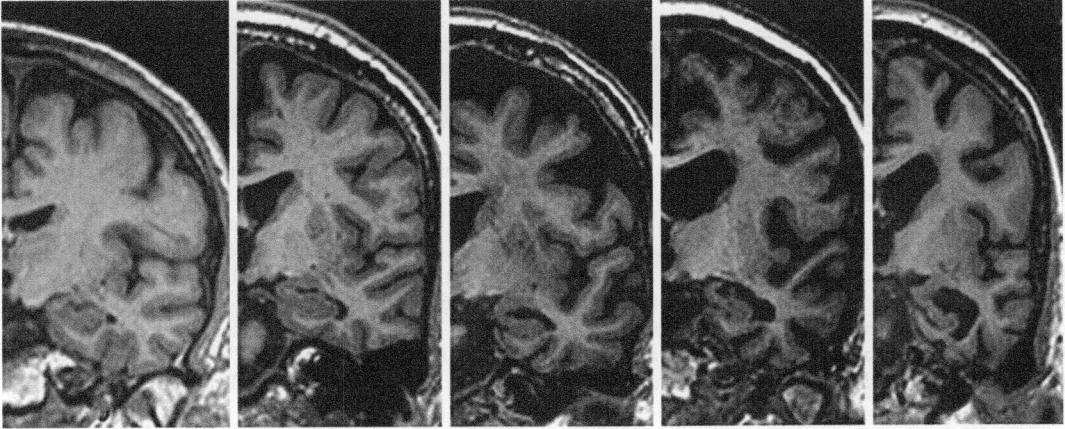

Fig. 1.7 Visual rating scale. MR images depict four degrees of atrophy in the hippocampus (HP) and entorhinal cortex (ERC) according to a visual rating scale in which 0 = no atrophy, 1 = minimal atrophy, 2 = mild atrophy, 3 = moderate atrophy, and 4 = severe atrophy. (Score shown corresponds to both structures.) (From Duara et al. 2013)

1.3.2 FDG PET

The well-established tracer [18F]Fluorodeoxy-glucose (FDG) allows the PET imaging of cerebral glucose metabolism, known to be tightly associated with neuronal function. Typical patterns of hypometabolism have been described in AD, including posterior parietal regions, precuneus, and frontal cortical regions, sparing sensorimotor and visual cortex. Numerous studies have demonstrated that early abnormalities, particularly in the posterior cingulate and precuneus cortical regions, have high positive and negative predictive value with regard to conversion to AD in the stage of MCI (Fig. 1.10). Even in some individuals with subjective memory impairment, changes in metabolism have been observed, potentially reflecting early pathological changes in the brain typical of AD (Mosconi et al. 2010; Protas et al. 2013; Silverman et al. 1999). However, brain metabolism may also remain normal despite the presence of significant amyloid accumulation in early MCI (Wu et al. 2012) and discrimination between naMCI and aMCI is poor (Lowe et al. 2009).

1.3.3 Molecular Neuroimaging

PET is a robust marker of neocortical fibrillar amyloid deposition in brain (*see* Fig. 4.2) (Nordberg et al. 2013). However, there is wide individual variation in the brain amyloid load in

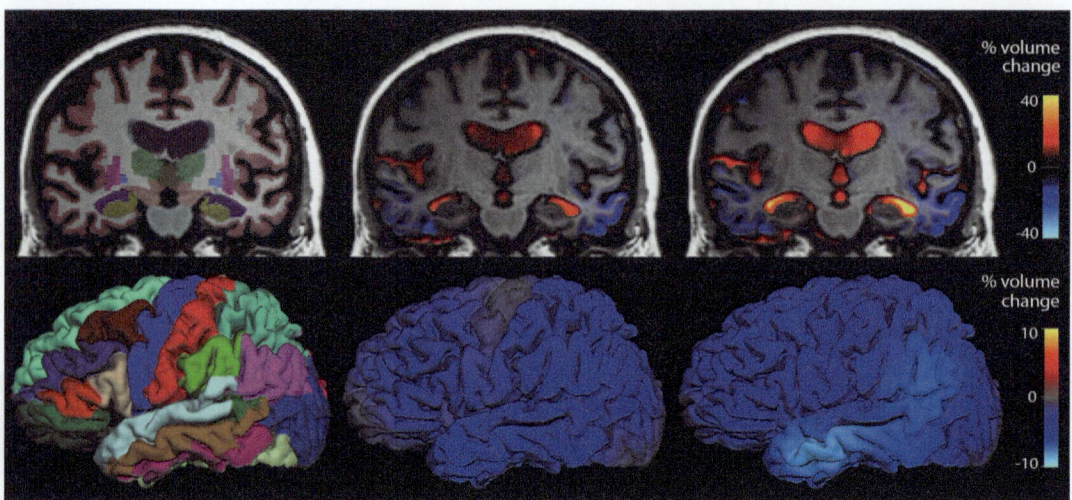

Fig. 1.8 Tissue segmentation, with 6-month and 12-month volume change fields for an individual with MCI. Segmentation of the baseline MRI scan, with different brain structures represented in different colors (*top left*). Corresponding coronal slice overlaid with a heat-map representation of the voxelwise estimates of volumetric changes at 6 and 12 months (*top middle* and *right*). Left hemisphere cortical parcellation of the baseline MRI scan (*bottom left*). Cortical surface overlaid with a heat-map representation of the estimates of cortical volumetric change at 6 months and 12 months (*bottom middle and right*). Region-specific estimates were obtained by averaging the voxelwise change within each region of interest. In this subject, the left middle-temporal gyrus has decreased in volume by 4.7 % at 6 months and by 8.2 % at 12 months; the left temporal-horn lateral ventricle has increased in volume by 17.4 % at 6 months and by 35.3 % at 12 months. (From Holland; with permission)

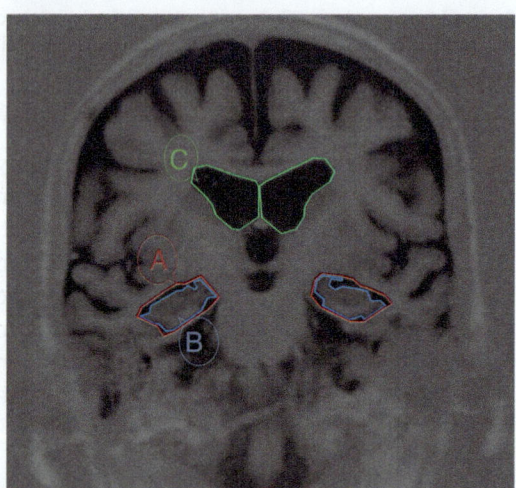

Fig. 1.9 MRI showing the areas needed to calculate the Medial Temporal Lobe Atrophy index (MTAi). MTAi = (A − B) × 10/C. **a** is the medial temporal lobe region; **b** is the parenchyma within the medial temporal region, including the hippocampus and the parahippocampal gyrus; and **c** is the body of the ipsilateral lateral ventricle (from Menéndez-González et al.)

MCI (Kemppainen et al. 2013). Several tracers for PET amyloid imaging have been evaluated successfully. At present, the first widely researched imaging technique was carbon 11–labeled Pittsburgh Compound B ($[^{11}C]$PiB) to detect amyloid plaques, but the short half-life of carbon 11 (approximately 20.4 min) limits its clinical application. On the other hand, PET tracers utilizing radiolabeled fluorine-18 have been designed to overcome this limitation; the longer half-life of this isotope (about 109.4 min) makes off-site preparation and regional distribution easier. Today, both Florbetapir and Flutemetamol have been approved by the US Food and Drug Administration (FDA) to evaluate adults for AD and other causes of cognitive decline.

A distinct uptake of the amyloid tracers has been observed in AD patients throughout the cerebral cortex, including frontal and temporoparietal regions and the precuneus (Huang et al. 2013) (Fig. 1.11). Amyloid deposition occurs relatively early in the precuneus, frontal region,

Table 1.2 Volumetric indices for comparing the extent/rate of atrophy in the medial temporal lobe with the extent/rate of global brain atrophy

	Compares	Parameters needed to calculate it	Computing	Interpretation
Temporal horn index (Ti)	Volume of the temporal horn with the volume of the lateral ventricles	Temporal horn volume (A) and the lateral ventricular volume (B)	$THi = A / B$	Low values are suggestive of MTL atrophy, and therefore the pattern of atrophy matches the pattern expected in typical AD
Medial Temporal-Lobe ratio (MTLr)	Volume of the MTL with the ipsilateral hemispheric volume	The volume of the hippocampus (A); the volume of the parahippocampal gyrus (B); the volume of the whole brain hemisphere (C)	$MTLr = (A+B)^2 / C$	Low values are suggestive of MTL atrophy, and therefore the pattern of atrophy matches the pattern expected in typical AD
Yearly rate of MTL atrophy (yrMTA)	Not an index	A and B as in MTLr in 2 different MRI studies	$(yrMTL) = (A1+B1) - (A2+B2) \cdot 1200 / (\text{# mo between MRI studies})$	High values are expected in typical AD
Yearly rate of relative MTL atrophy (yrRMTA)	Rate of atrophy of the MTL with the rate of enlargement of the ipsilateral lateral ventricles	A, B and C as in MTLr in 2 different MRI studies	$yrrMTL = (A1+B1) - (A2+B2) \cdot 1200 / (C2-C1) \cdot (\text{# mo between MRI studies})$	High values are expected in early typical AD
Hippocampus ratio (Hr)	Volume of the hippocampus with the ipsilateral hemispheric volume	The volume of the hippocampus (A); the volume of the ipsilateral brain hemisphere (B)	$Hr = A^2 / B$	Low values are suggestive of hippocampus atrophy, and therefore the pattern of atrophy matches the pattern expected in typical AD
Yearly rate of Hippocampus Atrophy (yrHA)	Not an index	A as in Hr in 2 different MRI studies	$(yrHA) = (A1-A2) \cdot 1200 / (\text{# mo between MRI studies})$	High values are expected in typical AD
Yearly rate of relative Hippocampus Atrophy (yrRHA)	Rate of atrophy of the hippocampus with the rate of atrophy of the ipsilateral hemisphere	A and B as in Hr in 2 different MRI studies	$yrrHA = (A1-A2) \cdot 1200 / (B1-B2) \cdot (\text{# mo between MRI studies})$	High values are expected in early typical AD
Hippocampus-Ventricle index (HVi)	Addition of the volume of the hippocampus plus the 10th part of the volume of the lateral ventricle	Normalized volume of the hippocampus (A); normalized volume of the lateral ventricle (B)	$HVi = A + (B / 10)$	Low HVi values are suggestive of AD pathology in incipient stages; high HVi values are suggestive of global brain atrophy due to aging or any neurodegenerative disease other than AD

AD—Alzheimer's disease; MTL—medial temporal lobe

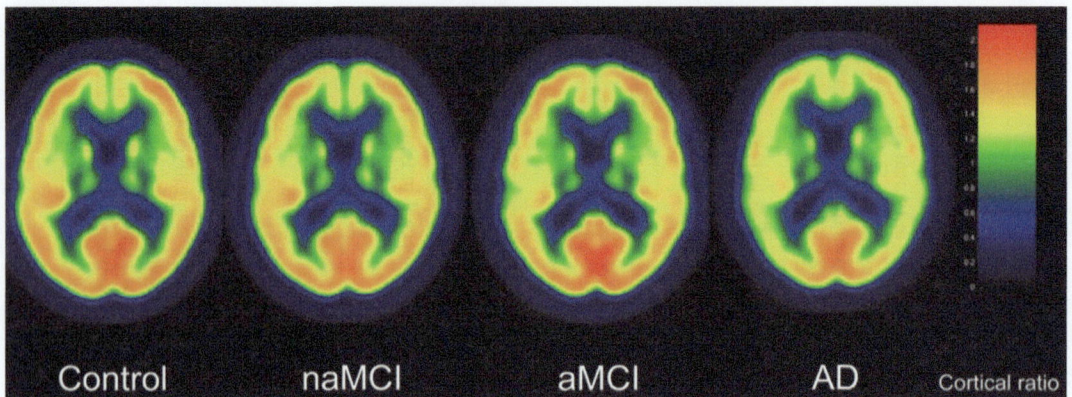

Fig. 1.10 FDG-PET images for control, non-amnestic MCI (naMCI), amnestic MCI (aMCI), and AD subjects. In the color scale of glucose metabolism, red is high, yellow is intermediate, and green is low. The levels of glucose metabolism in the brain are decreased in patients with AD. Discrimination between naMCI and aMCI is poor (from Lowe; with permission)

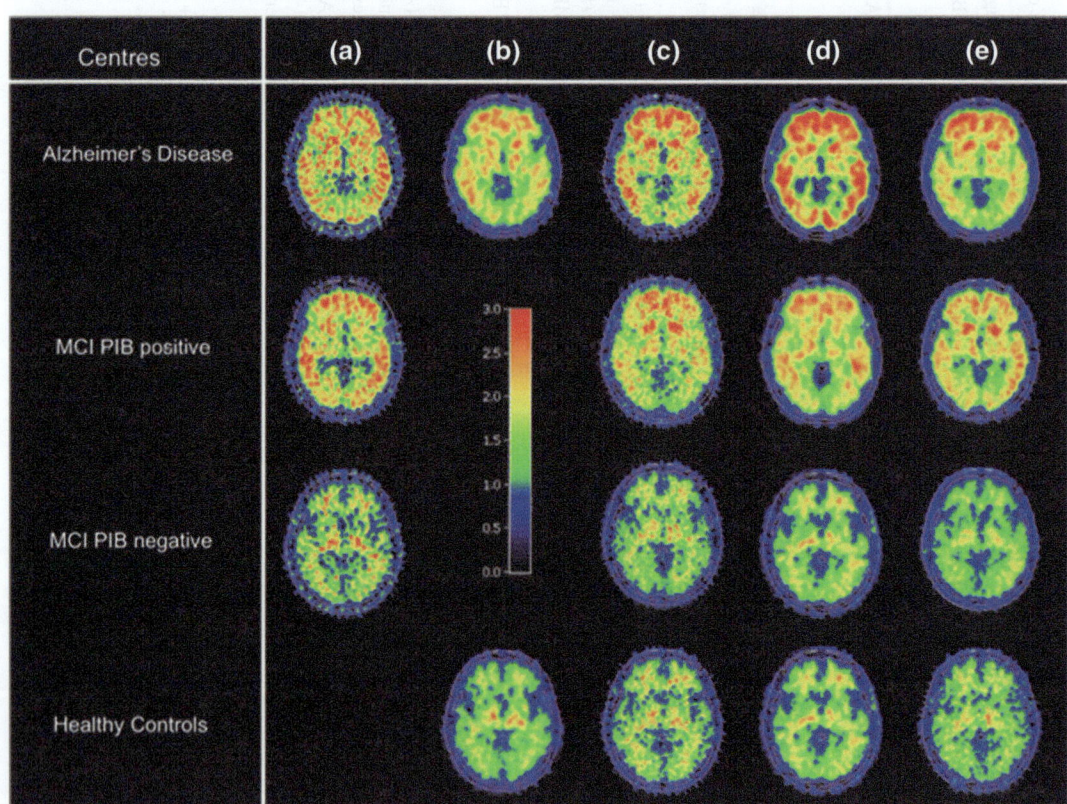

Fig. 1.11 Representative sagittal Florbetapir images. In the *bottom row*, varying degrees of [18F]AV-45 PET uptake can be seen in cognitively normal (CN) subjects. In the *middle rows*, subjects with amnestic MCI (aMCI) negative or positive for cerebral amyloidosis, and subjects with AD. In the *top row*, the uptake closely corresponds to the pathological amyloid deposition distribution as described by Braak and Braak; in the color scale of amyloid accumulation, red is high, green is intermediate, and blue is low (from Huang et al. 2013)

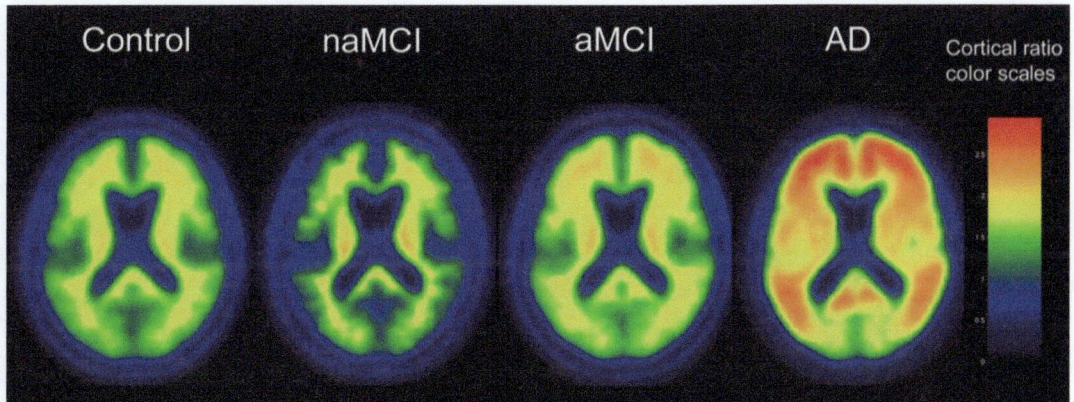

Fig. 1.12 PIB-PET images for control, non-amnestic MCI (naMCI), amnestic MCI (aMCI), and AD subjects. In the color scale of amyloid uptake, *red* is high, *yellow* is intermediate, and *green* is low. The levels of amyloid uptake in the brain are increased in the subject with aMCI and specially in the subject with AD. Discrimination between naMCI and aMCI is good (from Lowe; with permission)

and posterior cingulate in subjects with amnestic MCI (aMCI). Whereas the basal ganglia are also regularly affected, sensorimotor and visual cortical regions show less uptake, and the cerebellum is free of any relevant gray-matter tracer accumulation. Significant correlation between Mini-Mental State Examination (MMSE) scores and amyloid accumulation can be observed among normal controls and subjects with aMCI and AD. In addition, discrimination between naMCI and aMCI is good (Lowe et al. 2009).

In vivo versus postmortem histopathological cross-evaluation studies have also confirmed that increased cortical tracer uptake corresponds to amyloid aggregation in the brain. In young, healthy control subjects, no gray-matter binding of the amyloid tracers is observed; only non-specific tracer uptake in the white matter has been demonstrated. In general, this white-matter uptake has been described to be less pronounced for [^{11}C]PiB than for the 18F-labeled compounds, which may somewhat decrease the sensitivity of the 18F-labeled versions of amyloid tracers (Fig. 1.12).

With regard to early diagnosis, a number of studies have demonstrated a high predictive value of a positive amyloid scan in the MCI stage with regard to conversion to AD (Fig. 1.13).

Even in subjects with SCI, increased levels of amyloid deposition have been described.

Researchers were also able to demonstrate elevated amyloid levels in asymptomatic carriers of the APOE ε4 allele (Reiman et al. 2009). Furthermore, in a relevant proportion of elderly subjects without any cognitive complaints, elevated cortical tracer uptake was observed consistently. The meaning of these observations is not definitely clear so far, but a number of findings indicate that these subjects may indeed suffer from early AD pathology, potentially leading to dementia later in life. These findings include relatively worse performance in cognitive tests and abnormal findings in other imaging tests such as resting-state connectivity.

1.3.4 Key Concepts

1. Atrophy of MTL structures is now considered to be a valid diagnostic marker at the MCI stage.
2. Atrophy of the MTL can be assessed using different technologies, from simple visual rating scales to sophisticated volumetric methods.
3. The medial temporal lobe atrophy index (MTAi) is a simple and objective two-dimensional method for assessing MTL atrophy.

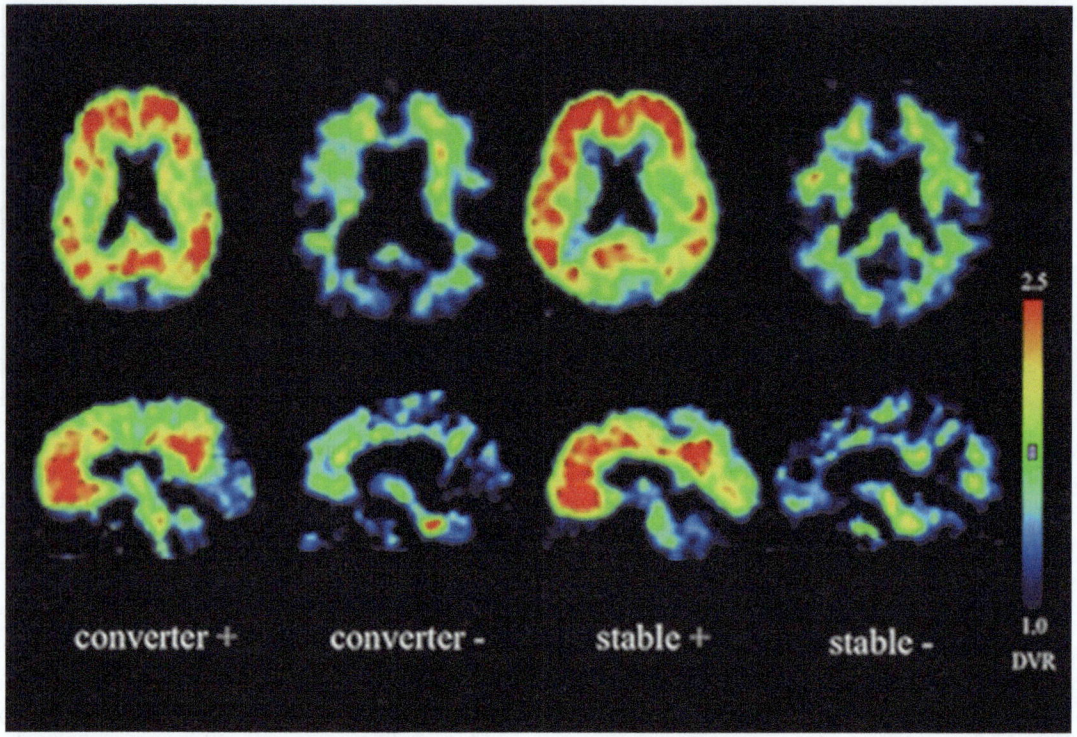

Fig. 1.13 PiB-PET distribution volume ratio (DVR) images from representative MCI converters and stable patients with (+) and without (−) amyloid deposition at baseline. (Hatashita and Yamasaki 2013). In the color scale of PiB retention, *red* is high, *green* is intermediate, and *blue* is low

4. Several indices are available to compare the extent of MTL atrophy versus global brain atrophy.
5. FDG PET shows a decrease in metabolic rate for glucose in the posterior cingulate, parietal, temporal, and prefrontal cortex in patients with MCI and AD.
6. A distinct uptake of the amyloid tracers has been observed in AD patients throughout the cerebral cortex, including frontal and temporoparietal regions and the precuneus. In vivo versus postmortem histopathological cross-evaluation studies have also confirmed that increased cortical tracer uptake corresponds to amyloid aggregation in the brain.
7. Studies have demonstrated that a positive amyloid scan in the stage of MCI has a high predictive value with regard to conversion to AD.

References

Blennow K (2004) Cerebrospinal fluid protein biomarkers for Alzheimer's disease. NeuroRx 1:213–225

Blennow K, Vanmechelen E, Hampel H (2001) CSF total tau, Abeta42 and phosphorylated tau protein as biomarkers for Alzheimer's disease. Mol Neurobiol 24:87–97

Buchhave P, Blennow K, Zetterberg H, Stomrud E, Londos E et al (2009) Longitudinal study of CSF biomarkers in patients with Alzheimer's disease. PLoS ONE 4:e6294

Buchhave P, Minthon L, Zetterberg H, Wallin AK, Blennow K, Hansson O (2012) Cerebrospinal fluid levels of β-amyloid 1-42, but not of tau, are fully changed already 5 to 10 years before the onset of Alzheimer dementia. Arch Gen Psychiatry 69:98–106

Chételat G, Villemagne VL, Villain N, Jones G, Ellis KA, Ames D et al (2012) Accelerated cortical atrophy in cognitively normal elderly with high β-amyloid deposition. Neurology 78:477–484

Duara R, Loewenstein DA, Shen Q, Barker W, Varon D, Greig MT et al (2013) The utility of age-specific cutoffs for visual rating of medial temporal atrophy in classifying Alzheimer's disease, MCI and cognitively normal elderly subjects. Front Aging Neurosci 5:47

Forlenza OV, Diniz BS, Gattaz WF (2010) Diagnosis and biomarkers of predementia in Alzheimer's disease. BMC Med 8:89

Frankó E, Joly O (2013) Alzheimer's disease neuroimaging initiative. Evaluating Alzheimer's disease progression using rate of regional hippocampal atrophy. PLoS ONE 8:e71354

Giesel FL, Hahn HK, Thomann PA, Widjaja E, Wignall E, von Tengg-Kobligk H et al (2006) Temporal horn index and volume of medial temporal lobe atrophy using a new semiautomated method for rapid and precise assessment. AJNR Am J Neuroradiol 27:1454–1458

Hatashita S, Yamasaki H (2013) Diagnosed mild cognitive impairment due to Alzheimer's disease with PET biomarkers of beta amyloid and neuronal dysfunction. PLoS ONE 8:e66877

Holland D, Brewer JB, Hagler DJ, Fennema-Notestine C, Dale AM (2009) Alzheimer's disease neuroimaging initiative. Subregional neuroanatomical change as a biomarker for Alzheimer's disease. Proc Natl Acad Sci USA 106(49):20954–20959. [Erratum in: Proc Natl Acad Sci USA 107(14):6551 (2010)]

Huang KL, Lin KJ, Hsiao IT, Kuo HC, Hsu WC, Chuang WL et al (2013) Regional amyloid deposition in amnestic mild cognitive impairment and Alzheimer's disease evaluated by [18F]AV-45 positron emission tomography in Chinese population. PLoS ONE 8:e58974

Kemppainen NM, Scheinin NM, Koivunen J, Johansson J, Toivonen JT, Någren K et al (2014) Five-year follow-up of 11C-PIB uptake in Alzheimer's disease and MCI. Eur J Nucl Med Mol Imaging 41(2):283–289

Koepsell TD, Monsell SE (2012) Reversion from mild cognitive impairment to normal or near-normal cognition: risk factors and prognosis. Neurology 79: 1591–1598. doi:10.1212/WNL.0b013e31826e26b7

Lazarczyk MJ, Hof PR, Bouras C, Giannakopoulos P (2012) Preclinical Alzheimer disease: identification of cases at risk among cognitively intact older individuals. BMC Med 10:127

Lopez OL, Becker JT, Chang YF, Sweet RA, DeKosky ST, Gach MH et al (2012) Incidence of mild cognitive impairment in the Pittsburgh cardiovascular health study-cognition study. Neurology 79:1599–1606

Lowe VJ, Kemp BJ, Jack CR Jr, Senjem M, Weigand S, Shiung M, Smith G, Knopman D, Boeve B, Mullan B, Petersen RC (2009) Comparison of 18F-FDG and PiB PET in cognitive impairment. J Nucl Med 50(6):878–886

Menéndez-González M, López-Muñiz A, Vega JA, Salas-Pacheco JM, Arias-Carrión O (2014) MTA index: a simple 2D-method for assessing atrophy of the medial temporal lobe using clinically available neuroimaging. Front Aging Neurosci 6:23

Menéndez-González M (2014a) Volumetric indices and rates of atrophy for the assessment of medial temporal lobe atrophy. J Neurol Neurosci 5(2):1

Menéndez-González M (2014b) Routine lumbar puncture for the early diagnosis of Alzheimer's disease. Is it safe? Front Aging Neurosci 8:65

Mosconi L, Berti V, Glodzik L, Pupi A, De Santi S, de Leon MJ (2010) Pre-clinical detection of Alzheimer's disease using FDG-PET, with or without amyloid imaging. J Alzheimers Dis 20:843–854

Nordberg A, Carter SF, Rinne J, Drzezga A, Brooks DJ, Vandenberghe R et al (2013) A European multicentre PET study of fibrillar amyloid in Alzheimer's disease. Eur J Nucl Med Mol Imaging 40:104–114

Protas HD, Chen K, Langbaum JS, Fleisher AS, Alexander GE, Lee W et al (2013) Posterior cingulate glucose metabolism, hippocampal glucose metabolism, and hippocampal volume in cognitively normal, late-middle-aged persons at 3 levels of genetic risk for Alzheimer disease. JAMA Neurol 70:320–325

Quiroz YT, Stern CE, Reiman EM, Brickhouse M, Ruiz A, Sperling RA et al (2013) Cortical atrophy in presymptomatic Alzheimer's disease presenilin 1 mutation carriers. J Neurol Neurosurg Psychiatry 84:556–561

Reiman EM, Chen K, Liu X, Bandy D, Yu M, Lee W et al (2009) Fibrillar amyloid-beta burden in cognitively normal people at 3 levels of genetic risk for Alzheimer's disease. Proc Natl Acad Sci USA 106:6820–6825

Rodrigue KM, Kennedy KM, Devous MD Sr, Rieck JR, Hebrank AC, Diaz-Arrastia R et al (2012) β-Amyloid burden in healthy aging: regional distribution and cognitive consequences. Neurology 78:387–395

Scheef L, Spottke A, Daerr M, Joe A, Striepens N, Kölsch H et al (2012) Glucose metabolism, gray matter structure, and memory decline in subjective memory impairment. Neurology 79:1332–1339

Silverman DHS, Small GW, Phelps ME (1999) Clinical value of neuroimaging in the diagnosis of dementia: sensitivity and specificity of regional cerebral metabolic and other parameters for early identification of Alzheimer's disease. Clin Positron Imaging 2:119–130

van Harten AC, Visser PJ, Pijnenburg YA, Teunissen CE, Blankenstein MA, Scheltens P et al (2013) Cerebrospinal fluid Aβ42 is the best predictor of clinical progression in patients with subjective complaints. Alzheimers Dement 9:481–487

van Rossum IA, Vos SJ, Burns L, Knol DL, Scheltens P, Soininen H et al (2012) Injury markers predict time to dementia in subjects with MCI and amyloid pathology. Neurology 79:1809–1816

Westman E, Cavallin L, Muehlboeck JS, Zhang Y, Mecocci P, Vellas B et al (2011) Sensitivity and specificity of medial temporal lobe visual ratings and multivariate regional MRI classification in Alzheimer's disease. PLoS ONE 6:e22506

Wu L, Rowley J, Mohades S, Leuzy A, Dauar MT, Shin M et al (2012) Dissociation between brain amyloid deposition and metabolism in early mild cognitive impairment. PLoS ONE 7:e47905

2.1 Introduction

AD accounts for up to 75 % of all dementia cases. The differential diagnosis with other conditions is sometimes challenging since several disorders may produce similar symptoms to AD, including cortical basal degeneration (CBD), dementia with Lewy bodies (DLB), Parkinson's disease dementia (PDD), frontotemporal lobe dementia (FTLD), vascular dementia (VaD), multiple-system atrophy (MSA) and progressive supranuclear palsy (PSP), among others.

Knowing the key features and pathology of each type of dementia can help in accurate diagnosis, so that patients will receive the treatment and support services appropriate for their condition and maintain the highest possible quality of life (Table 2.1) (Seeley and Miller 2013).

However, biomarkers able to differential between dementing conditions would be very useful in clinical practice. Indeed, today there is evidence that some biomarkers can play a role in the differential diagnosis of AD against other dementias (Table 2.2).

2.2 CSF Biomarkers

CSF Aβ42, t-Tau, and p-Tau are useful in differential diagnosis of dementing disorders, including AD, DLB, FTLD, VaD, CBD, Creutzfeldt-Jakob disease, psychiatric disorders, and subjective memory complaints (SMC). One must be aware, however, that there is some overlapping in the proteinopathies behind the different neurodegenerative disorders, and this fact is translated in the level of CSF proteins. For instance, CSF Tau elevation may be observed also in other diseases, being implicated pathologically in PD, PSP, and CBD (Zhang et al. 2013), potentially limiting the utility of Tau alone in the differential diagnosis of dementia. Nevertheless, the p-Tau/Aβ42 ratio is a useful tool to discriminate AD from both FTLD and semantic dementia (SD) (de Souza et al. 2011). Regarding the different proteinopathies behind the most frequent neuropathological subtypes of FTLD (Tau and TDP), CSF measurements of Aβ1−42, t-Tau, and p-Tau in FTLD differ significantly from the abnormal levels seen in AD, and in a subset of both FTLD-Tau and FTLD-TDP there are extremely low levels of t-Tau of unclear etiology. These properties allow for accurate distinction of FTLD from AD in autopsy-confirmed cohorts, though FTLD-specific markers are still lacking (Irwin et al. 2013).

Synucleinopathies (PD, DLB, MSA) are caused by the accumulation of α-synuclein in the brain. CSF levels of α-synuclein are decreased in patients with PD, DLB, and MSA but increased in patients with AD; thus CSF α-synuclein differentiates these synucleinopathies from AD (Wennström et al. 2013; van Dijk et al. 2013).

M. Menéndez González, *Atlas of Biomarkers for Alzheimer's Disease*,
DOI: 10.1007/978-3-319-07989-9_2, © Springer International Publishing Switzerland 2014

Table 2.1 Clinical differentiation of the major dementias

Disease	First symptom	Mental status	Neuropsychiatry	Neurology	Imaging
AD	Memory loss	Episodic memory loss	Initially normal	Initially normal	Entorhinal cortex and hippocampal atrophy
FTLD	Apathy; poor judgment/insight, speech/language; hyperorality	Frontal/ executive, language; spares drawing	Apathy, disinhibition, hyperorality, euphoria, depression	May have vertical gaze palsy, axial rigidity, dystonia, alien hand, or MND	Frontal, insular, and/or temporal atrophy; spares posterior parietal lobe
DLB/ PDD	Visual hallucinations, REM sleep disorder, delirium, Capgras syndrome, parkinsonism	Drawing and frontal/ executive; spares memory; delirium prone	Visual hallucinations, depression, sleep disorder, delusions	Parkinsonism	Posterior parietal atrophy; hippocampi larger than in AD
Vascular dementia	Often but not always sudden; variable; apathy, falls, focal weakness	Frontal/ executive, cognitive slowing; can spare memory	Apathy, delusions, anxiety	Usually motor slowing, spasticity; can be normal	Cortical and/or subcortical infarctions, confluent white matter

AD Alzheimer's disease, *CBD* cortical basal degeneration, *DLB* dementia with Lewy bodies, *FTLD* frontotemporal lobe dementia, *MND* motor neuron disease, *PSP* progressive supranuclear palsy, *PDD* Parkinson's disease dementia, *REM* rapid eye movement

Table 2.2 Biomarkers potentially useful in the differential diagnosis of AD against other dementias

Other dementia	Useful biomarkers
Parkinson's disease dementia	CSF α-synuclein, amyloid
Dementia with Lewy bodies	Medial temporal atrophy on MRI, CSF α-synuclein
Frontotemporal lobe dementia[a]	CSF p-Tau/Aβ42 ratio, medial temporal atrophy on MRI, amyloid, FDG PET
Multiple-system atrophy	CSF α-synuclein
Primary progressive apraxia	Medial temporal atrophy on MRI
Vascular dementia	Amyloid, CSF p-Tau/Aβ42 ratio

[a] The three clinical subtypes show different patterns of hippocampal atrophy: (1) Semantic dementia shows bilateral hippocampal atrophy, although the left hippocampus tends to be smaller than in AD. (2) The behavioral variant (bvFTD) group showed significant white matter contraction, and the presence of behavioral symptoms at baseline predicted significant volume loss of the ventromedial prefrontal cortex (Lu et al. 2013). (3) No significant hippocampal atrophy was detected in nonfluent progressive aphasia (van de Pol et al. 2006)

CSF cerebrospinal fluid, *FDG PET* fluorodeoxyglucose positron emission tomography, *Amyloid* represents both CSF amyloid and amyloid PET SCAN

2.3 Neuroimaging

2.3.1 MRI

By highlighting specific topographical patterns of atrophy, neuroimaging using MRI has the potential to be useful in discriminating between different types of dementia. Presently, scanner manufacturers are developing radiological expert systems based on indices and ratios to help in rating the presence or absence of AD from the pattern of atrophy derived from a single brain scan. The major impetus for the development of methods of temporal lobe assessment has been to improve the accuracy of early diagnosis in AD, yet it is unclear whether medial temporal lobe atrophy is specific for AD or can also be a feature of other dementias, as few studies to date have compared patients with AD with patients having other forms of degenerative disease, such as FTLD.

Several studies support the hypothesis that a greater burden of pathology centers on the temporal lobes in AD compared with DLB, except in DLB cases with concurrent AD pathology (Barber et al. 1999; Tam et al. 2005; Burton et al. 2009). Unlike findings reported in younger subjects, visual ratings for posterior cortical atrophy are not a reliable marker at older ages for distinguishing AD from controls, or for distinguishing DLB from AD (O'Donovan et al. 2013). However, visual ratings of medial temporal lobe atrophy (MTA) may be useful markers in distinguishing both AD and DLB in older patients without dementia. Similarly, more objective methods such as the MTA index (MTAi) or volumetric studies are expected to be useful in the differential diagnosis of AD against other neurodegenerative dementias in a wider range of ages.

2.3.2 FDG PET

Regarding differential diagnosis, [18F]fluorodeoxyglucose (FDG) PET has shown great value because it allows the detection of different patterns of neurodegeneration that are specific for various non-AD (amyloid-negative) forms of neurodegeneration. These include the subtypes of FTLD, as well as MSA, CBD, and PSP (Yamane et al. 2013). Most importantly, FDG PET is also highly useful in differentiating within amyloid-positive subtypes of disease, which cannot be distinguished on the basis of their amyloid PET scan. This includes Lewy-body dementia (Kasanuki et al. 2012; Lim et al. 2009; Ishii et al. 2007), posterior cortical atrophy, and the logopenic variant of progressive aphasia (Fig. 2.1) (Madhavan et al. 2013; Mosconi et al. 2008).

2.3.3 Molecular Neuroimaging

Most molecular tracers used in AD research are considered to be specific for amyloid deposition, with the exception of FDDNP, which has been demonstrated to bind also to Tau aggregates apart from amyloid. Although the tracers are specific for the protein aggregation (i.e., amyloid plaques), it is important to remember that the protein aggregation is not specific for AD. For example, it is known from histopathological studies that aggregation of amyloid plaques is found in the brain of most patients with DLB, in addition to the pathognomonic synuclein deposits. Thus, amyloid imaging may not be able to differentiate between DLB and AD. Furthermore, amyloid imaging alone may not be helpful with regard to distinguishing between amyloid-positive subtypes of AD (typical AD, the logopenic variant of progressive aphasia, and posterior cortical atrophy).

A negative amyloid-PET scan indicates few or no neuritic plaques and is inconsistent with a neuropathological diagnosis of AD at the time of image acquisition. Thus, it also reduces the likelihood that a patient's cognitive impairment is due to AD. On the other hand, a positive amyloid-PET scan indicates moderate to frequent amyloid neuritic plaques. Such scans may be observed in older people with normal cognition and in patients with various neurologic conditions, including AD. Therefore a positive scan alone does not establish a diagnosis of AD or other cognitive disorder, although it may be useful in combination with other clinical parameters. Pittsburgh Compound B (PiB) and FDG are similarly sensitive in discriminating between AD and FTLD. PiB is more sensitive when interpreted qualitatively or quantitatively. FDG is more specific, but only when scans are classified quantitatively (Rabinovici et al. 2011).

2.4 Key Concepts

1. There is some overlap among the proteins accumulated in various neurodegenerative disorders, and this fact is translated in the level of CSF proteins, limiting the utility of such measures in the differential diagnosis of dementia.
2. Several CSF proteins are useful in the differential diagnosis between the

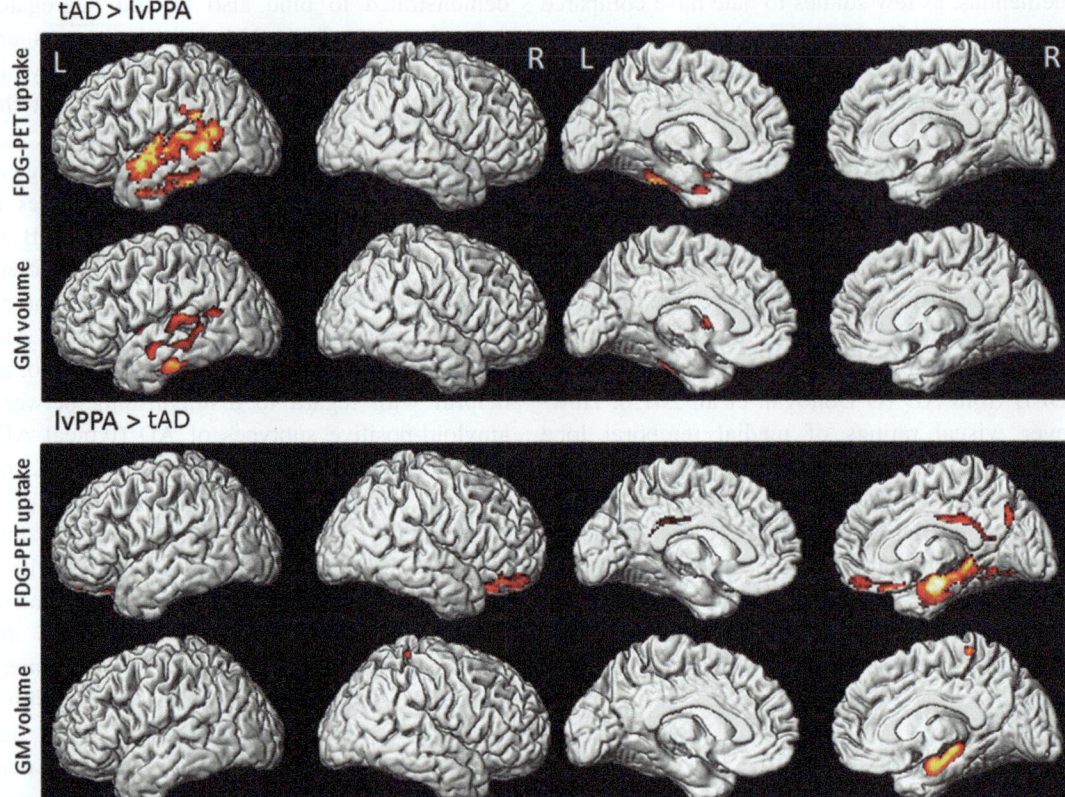

Fig. 2.1 Voxel-level imaging comparison of the logopenic variant of primary progressive aphasia (lvPPA)—an atypical clinical variant of Alzheimer's disease—and typical Alzheimer's type (tAD). A higher degree of atrophy was observed in lateral temporoparietal and medial parietal lobes, left greater than right, and left frontal lobe in the logopenic group (from Madhavan et al.)

neurodegenerative disorders causing dementia, however. For instance, the p-Tau/Aβ42 ratio is a useful tool to discriminate AD from FTLD.

3. CSF α-synuclein is useful in the differentiating AD from PDD, DLB, and MSA.
4. Hippocampal atrophy is not specific for AD, but it is useful to differentiate AD from DLB.
5. There are some differences between the profile of hippocampus atrophy found in AD and the profile found in the different types of FTLD.
6. A negative amyloid PET scan rules out AD with high predictive value.
7. PiB and FDG PET scans are useful in discriminating AD from FTLD. FDG PET is also useful for differentiating AD from DLB.

References

Barber R, Gholkar A, Scheltens P, Ballard C, McKeith IG, O'Brien JT (1999) Medial temporal lobe atrophy on MRI in dementia with Lewy bodies. Neurology 52:1153–1158

Burton EJ, Barber R, Mukaetova-Ladinska EB, Robson J, Perry RH, Jaros E et al (2009) Medial temporal lobe atrophy on MRI differentiates Alzheimer's disease from dementia with Lewy bodies and vascular cognitive impairment: a prospective study with pathological verification of diagnosis. Brain 132(Pt 1):195–203

de Souza LC, Lamari F, Belliard S, Jardel C, Houillier C, De Paz R et al (2011) Cerebrospinal fluid biomarkers in the differential diagnosis of Alzheimer's disease from other cortical dementias. J Neurol Neurosurg Psychiatry 82:240–246

Irwin DJ, Trojanowski JQ, Grossman M (2013) Cerebrospinal fluid biomarkers for differentiation of frontotemporal lobar degeneration from Alzheimer's disease. Front Aging Neurosci 5:6

Ishii K, Soma T, Kono AK, Sofue K, Miyamoto N, Yoshikawa T et al (2007) Comparison of regional brain volume and glucose metabolism between patients with mild dementia with Lewy bodies and those with mild Alzheimer's disease. J Nucl Med 48:704–711

Kasanuki K, Iseki E, Fujishiro H, Yamamoto R, Higashi S, Minegishi M et al (2012) Neuropathological investigation of the hypometabolic regions on positron emission tomography with [18F] fluorodeoxyglucose in patients with dementia with Lewy bodies. J Neurol Sci 314:111–119

Lim SM, Katsifis A, Villemagne VL, Best R, Jones G, Saling M et al (2009) The 18F-FDG PET cingulate island sign and comparison to 123I-beta-CIT SPECT for diagnosis of dementia with Lewy bodies. J Nucl Med 50:1638–1645

Lu PH, Mendez MF, Lee GJ, Leow AD, Lee HW, Shapira J et al (2013) Patterns of brain atrophy in clinical variants of frontotemporal lobar degeneration. Dement Geriatr Cogn Disord 35:34–50

Madhavan A, Whitwell JL, Weigand SD, Duffy JR, Strand EA, Machulda MM et al (2013) FDG PET and MRI in logopenic primary progressive aphasia versus dementia of the Alzheimer's type. PLoS One 8:e62471

Mosconi L, Tsui WH, Herholz K, Pupi A, Drzezga A, Lucignani G et al (2008) Multicenter standardized 18F-FDG PET diagnosis of mild cognitive impairment, Alzheimer's disease, and other dementias. J Nucl Med 49:390–398

O'Donovan J, Watson R, Colloby SJ, Firbank MJ, Burton EJ, Barber R et al (2013) Does posterior cortical atrophy on MRI discriminate between Alzheimer's disease, dementia with Lewy bodies, and normal aging? Int Psychogeriatr 25:111–119

Rabinovici GD, Rosen HJ, Alkalay A, Kornak J, Furst AJ, Agarwal N et al (2011) Amyloid vs FDG-PET in the differential diagnosis of AD and FTLD. Neurology 77:2034–2042

Seeley WW, Miller BL (2013) Alzheimer's disease and other dementias. In: Hauser SL (ed) Harrison's neurology in clinical medicine, 3rd edn. McGraw-Hill Education, New York, pp 310–332

Tam CW, Burton EJ, McKeith IG, Burn DJ, O'Brien JT (2005) Temporal lobe atrophy on MRI in Parkinson disease with dementia: a comparison with Alzheimer disease and dementia with Lewy bodies. Neurology 64:861–865

van de Pol LA, Hensel A, van der Flier WM, Visser PJ, Pijnenburg YA, Barkhof F, Gertz HJ et al (2006) Hippocampal atrophy on MRI in frontotemporal lobar degeneration and Alzheimer's disease. J Neurol Neurosurg Psych 77:439–442

van Dijk KD, Bidinosti M, Weiss A, Raijmakers P, Berendse HW, van de Berg WD (2013) Reduced α-synuclein levels in cerebrospinal fluid in Parkinson's disease are unrelated to clinical and imaging measures of disease severity. Eur J Neurol. doi:10.1111/ene.12176

Wennström M, Surova Y, Hall S, Nilsson C, Minthon L, Boström F et al (2013) Low CSF levels of both α-synuclein and the α-synuclein cleaving enzyme neurosin in patients with synucleinopathy. PLoS One 8:e53250

Yamane T, Ikari Y, Nishio T, Ishii K, Ishii K, Kato T et al (2013) Visual-statistical interpretation of 18F-FDG-PET images for characteristic Alzheimer patterns in a multicenter study: inter-rater concordance and relationship to automated quantitative evaluation. Am J Neuroradiol. doi:10.3174/ajnr.A3665

Zhang J, Mattison HA, Liu C, Ginghina C, Auinger P, McDermott MP et al (2013) Longitudinal assessment of tau and amyloid beta in cerebrospinal fluid of Parkinson disease. Acta Neuropathol 126:671–678

3.1 Introduction

The pathological processes of AD occur in sequence, starting with amyloid beta (Aβ) deposition and progressing into neurodegeneration through neuronal loss and synaptic dysfunction. Great consideration has been given to the hypothetical model for the sequence of pathologic events in AD suggested by Jack and colleagues, according to which biomarkers reflecting Aβ pathology become positive before those reproducing neuronal degeneration and tangle development (Jack et al. 2009). In other words, the modified Aβ metabolism seems to precede Tau-related disease and neuronal degeneration.

The notion that AD biomarkers become abnormal not simultaneously but rather in an ordered manner is crucial: It implies that certain biomarkers may be differentially sensitive at different stages in the disease. For instance, clinical trials could use biomarkers to select a more homogeneous cohort of patients and to measure the effects of disease-modifying drugs (Tarawneh and Holtzman 2010; Menéndez-González et al. 2011).

The dynamic AD biomarker model proposes that Aβ deposition becomes abnormal early in the course of the disease, followed next by biomarkers of Tau-mediated neuronal injury and dysfunction, and lastly by biomarkers of overt neurodegeneration. All these changes may be detectable prior to cognitive and functional impairment (Jack et al. 2009) (Fig. 3.1).

3.2 CSF Biomarkers

Low CSF Aβ42 (as well as amyloid PET) is a biomarker of brain Aβ plaque deposition. CSF Aβ42 starts dropping early but drops very slowly. Aβ42 levels do not correlate well with disease duration or severity, so it is not a good parameter to monitor the progression of the disease. In addition, changes in CSF Aβ42 vary over time, partly due to the inconsistency of the currently available CSF Aβ42 assays (Riverol and López 2011).

High CSF Tau is an indicator of Tau pathological changes and associated neuronal injury. Thus, CSF Tau shoots up when symptoms start. The slight increase in Tau over time observed in patients with AD is modest when compared with the relatively large difference in absolute Tau levels between AD patients and controls. Tau levels seem to remain relatively stable throughout the disease process and do not correlate with dementia severity. Therefore, CSF Tau is not useful as a state marker over time, nor does it seem promising as a surrogate marker for treatment efficacy in clinical trials (Tarawneh and Holtzman 2010; Riverol and López 2011).

In conclusion, CSF biomarkers are useful to detect changes in amyloid metabolism (CSF Aβ42) and neuronal injury (CSF Tau), but they are not useful to track the disease over time. However, the use of multiple longitudinal CSF specimens may be necessary to detect the time point at which CSF biomarkers convert from

M. Menéndez González, *Atlas of Biomarkers for Alzheimer's Disease*,
DOI: 10.1007/978-3-319-07989-9_3, © Springer International Publishing Switzerland 2014

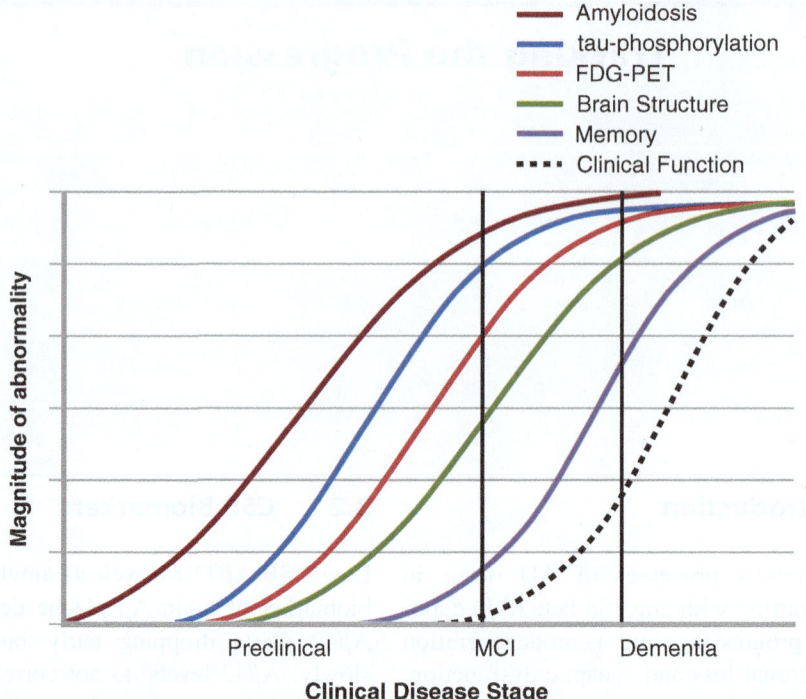

Fig. 3.1 The dynamic biomarkers of Alzheimer's disease (AD), as proposed by Jack and collaborators. The curves indicate changes caused by six studied biomarkers: amyloid β imaging detected in CSF and PET amyloid imaging (*brown line*); neurodegeneration detected by rise of CSF Tau (*blue line*); metabolic dysfunction measured with FDG PET (*red line*); brain atrophy and neuron loss measured with MRI (*green line*); cognitive decline measured by cognitive assessment (*purple line*) and loss of clinical function (*dotted line*). *MCI* mild cognitive impairment (adapted from Jack et al. 2009)

physiologic to pathologic values in some patients.

3.3   Neuroimaging Biomarkers

3.3.1   MRI

The hypothesis that cortical atrophy is present in preclinical AD more than 5 years prior to symptom onset is of crucial importance for the management of mild cognitive impairment (MCI). On MRI, the cortical thickness biomarker predicts cognitive decline in normal adults. Thus, this MRI biomarker may provide investigators with a population enriched for AD pathobiology and with a relatively high likelihood of imminent cognitive decline consistent with prodromal AD.

Medial temporal lobe atrophy (MTA) can be quantified using several different neuroimaging measurements, from rating scales to volumetric methods. The visual assessment rating scales shown in Chap. 1 (Fig. 1.7, Table 1.1) have failed to detect progression over time (Duara et al. 2013; Westman et al. 2011).

Researchers found that the pattern of hippocampus atrophy is similar in AD, MCI, and elderly controls, but the rate is significantly higher in patients with AD than in control subjects. Higher atrophy rates are also found in patients with progressive MCI compared with those having stable MCI, particularly in the anterolateral portion of the right hippocampus. Importantly, the regions showing the highest atrophy rate correspond to those that were described as having the highest burden of Tau deposition. Therefore local hippocampal atrophy

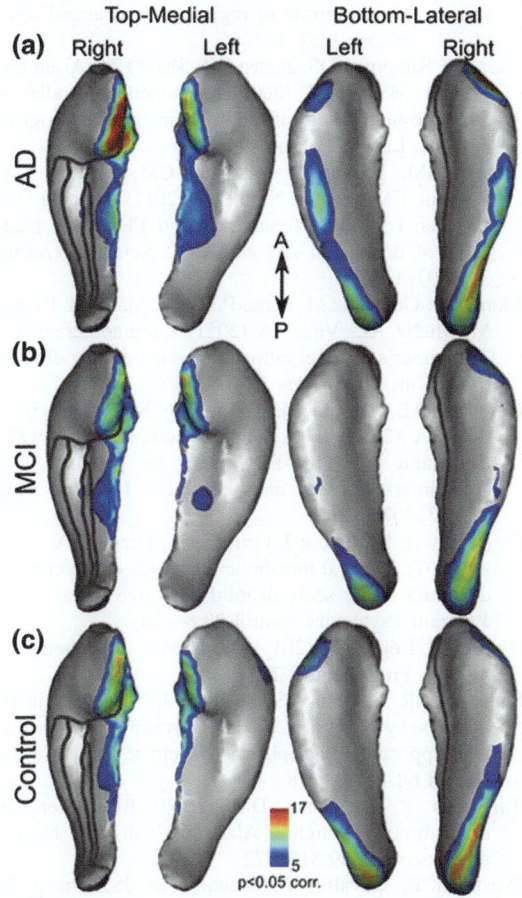

Fig. 3.2 Rates of significant hippocampal atrophy in patients with AD or MCI and controls, from a top-medial view (*first column*) and bottom-lateral view (*second column*) (from Frankó and Joly 2013)

rate is a reliable biomarker of disease stage and progression and could be considered as a method to objectively evaluate treatment effects (Frankó and Joly 2013) (Fig. 3.2). AD-specific cortical thinning and hippocampal volume loss are consistent with a sigmoidal pattern, with an acceleration phase during the early stages of the disease and a slower phase later (Sabuncu et al. 2011).

3.3.2 FDG PET

As noted in Chap. 1, hypometabolism on 18F-FDG PET is an indicator of impaired synaptic function accompanying neuronal injury. FDG PET has consistently shown a tight correlation between the level of metabolic decline and the degree of cognitive impairment (Perneczky et al. 2007; Newberg et al. 2012; Chen et al. 2011; Choo et al. 2007; Landau et al. 2011). This method thus qualifies for use in follow-up and therapy control studies. This correlation can be somewhat influenced by cognitive reserve effects, however, expressed in variable magnitude.

3.3.3 Molecular Neuroimaging

Researchers were able to demonstrate cerebral amyloid deposition using amyloid PET decades ahead of the expected onset of disease in autosomal dominantly inherited forms of AD. However, the expected time to a potential conversion to AD cannot be estimated on the basis of a positive amyloid scan alone. Furthermore, amyloid imaging is not suitable to detect soluble amyloid oligomers, which may represent the most toxic species. Only a limited correlation has been observed between in vivo measured amyloid burden and cognitive decline. This limitation may particularly depend on the stage of disease: In cognitively healthy elderly subjects, amyloid pathology may not yet have induced sufficient neurotoxic effects downstream from amyloid aggregation to have an impact on cognition, whereas in patients with manifest Alzheimer's dementia, a plateau of amyloid deposition has been observed, indicating that amyloid deposition reaches saturation while neurodegeneration (and cognitive decline) continues. Also, cerebral compensation mechanisms, expressed to different degrees in different subjects, may lead to a discrepancy between cortical amyloid load and symptomatic appearance.

3.4 Key Concepts

1. The longitudinal patterns support a hypothetical sequence of AD pathology in which amyloid deposition is an early event before hypometabolism or

hippocampal atrophy, suggesting that biomarker prediction of cognitive change is stage-dependent. Similarly, trajectories of the Aβ42 level in CSF, FDG uptake, and hippocampal volume vary across different cognitive stages.

2. Markers of brain Aβ plaque deposition, such as CSF Aβ, glucose metabolism, or amyloid PET, change early in the course of AD, whereas markers of neuronal injury and dysfunction, such as CSF Tau, hippocampus atrophy, or low glucose metabolism, change later.

3. Neither CSF Tau nor CSF Aβ42 are useful as state markers over time, nor do they seem promising as surrogate markers for treatment efficacy in clinical trials.

4. The local hippocampal atrophy rate is a reliable biomarker of disease stage and progression and could also be considered as a method to objectively evaluate treatment effects.

5. FDG PET findings correlate with the severity and progression of AD.

6. Amyloid PET shows a discrepancy between cortical amyloid load and symptomatic appearance.

References

Chen K, Ayutyanont N, Langbaum JB, Fleisher AS, Reschke C, Lee W et al (2011) Characterizing Alzheimer's disease using a hypometabolic convergence index. Neuroimage 56:52–60

Choo IH, Lee DY, Youn JC, Jhoo JH, Kim KW, Lee DS et al (2007) Topographic patterns of brain functional impairment progression according to clinical severity staging in 116 Alzheimer disease patients: FDG-PET study. Alzheimer Dis Assoc Disord 21:77–84

Duara R, Loewenstein DA, Shen Q, Barker W, Varon D, Greig MT et al (2013) The utility of age-specific cut-offs for visual rating of medial temporal atrophy in classifying Alzheimer's disease, MCI and cognitively normal elderly subjects. Front Aging Neurosci 5:47

Frankó E, Joly O (2013) Alzheimer's disease neuroimaging initiative. Evaluating Alzheimer's disease progression using rate of regional hippocampal atrophy. PLoS One 8:e71354

Jack CR, Knopman DS, Jagust WJ, Shaw LM, Aisen PS, Weiner MW et al (2009) Hypothetical model of dynamic biomarkers of the Alzheimer's pathological cascade. Lancet Neurol 9:119–128

Landau SM, Harvey D, Madison CM, Koeppe RA, Reiman EM, Foster NL et al (2011) Associations between cognitive, functional, and FDG-PET measures of decline in AD and MCI. Neurobiol Aging 32:1207–1218

Menéndez-González M, Pérez-Piñera P, Martínez-Rivera M, Muñiz AL, Vega JA (2011) Immunotherapy for Alzheimer's disease: rational basis in ongoing clinical trials. Curr Pharm Des 17:508–520

Newberg AB, Arnold SE, Wintering N, Rovner BW, Alavi A (2012) Initial clinical comparison of 18F-florbetapir and 18F-FDG PET in patients with Alzheimer disease and controls. J Nucl Med 53:902–907

Perneczky R, Hartmann J, Grimmer T, Drzezga A, Kurz A (2007) Cerebral metabolic correlates of the clinical dementia rating scale in mild cognitive impairment. J Geriatr Psychiatry Neurol 20:84–88

Riverol M, López OL (2011) Biomarkers in Alzheimer's disease. Front Neurol 2:46

Sabuncu MR, Desikan RS, Sepulcre J, Yeo BT, Liu H, Schmansky NJ et al (2011) The dynamics of cortical and hippocampal atrophy in Alzheimer disease. Arch Neurol 68:1040–1048

Tarawneh R, Holtzman DM (2010) Biomarkers in translational research of Alzheimer's disease. Neuropharmacology 59:310–322

Westman E, Cavallin L, Muehlboeck JS, Zhang Y, Mecocci P, Vellas B et al (2011) Sensitivity and specificity of medial temporal lobe visual ratings and multivariate regional MRI classification in Alzheimer's disease. PLoS One 6:e22506

Application of Alzheimer Biomarkers in Clinical Practice

4

4.1 Introduction

The research on biomarkers in the diagnosis of AD and its prodromal stage have created a need to translate research findings into tools for use in everyday clinical practice. Once a biomarker has proven validity, it should be integrated in clinical decision-making protocols. This is not an easy task, because many factors need to be taken into account and properly weighted (Fig. 4.1). In addition, the protocols may vary among health systems, so probably no protocol is valid for worldwide implementation. However, general recommendations for clinical practice should be reached within next few years.

Two basic questions are always raised when biomarkers are proposed for use in diagnosing cognitive disorders in clinical practice (Wilner 2010):

(1) Do biomarkers enhance diagnostic accuracy?

(2) Does the additional diagnostic accuracy provided by biomarkers really matter?

Clinical diagnostic accuracy for AD varies depending on the stage and the criteria used, but it generally is about 75–85 %. Today, there is consensus that several biomarkers, combined with the traditional clinical process, may allow a more accurate diagnosis for many dementias (Galluzi et al. 2013). This fact is particularly important in early stages, when the diagnosis is especially challenging. For most patients with

mild cognitive impairment (MCI), the standard practice used to include a careful history; physical, neurological, and neuropsychological evaluation; and close follow-up—a "wait and see" approach. Today we can offer a more proactive approach for discerning whether an underlying neurodegenerative process is behind the patient's mild deficits (Heister et al. 2011).

Does this additional diagnostic accuracy really matter? In clinical practice, a test that confirmed or ruled out AD would remove uncertainty. It would also clarify whether we should consider other diseases that may present symptoms similar to AD. Prompt diagnosis of some of these diseases can lead to earlier effective treatment, such as shunting for normal pressure hydrocephalus, supplementation with thyroid hormones in hypothyroidism, or antidepressive medications for depression.

The best argument for using biomarkers in clinical practice would be the possibility of treating patients with a disease-modifying therapy (such as drugs with neuroprotective effect) that prevents or delays the progression of the disease. But no drug has yet been proven to prevent AD. At best, the current therapies provide only symptomatic improvement, so there is an urgent need to discover neuroprotective treatments. But how can we conduct clinical trials to test such drugs when diagnosing AD at its very early stages is so difficult? Again, the support of biomarkers should be mandatory for enrolling patients in research studies.

M. Menéndez González, *Atlas of Biomarkers for Alzheimer's Disease*, DOI: 10.1007/978-3-319-07989-9_4, © Springer International Publishing Switzerland 2014

Fig. 4.1 Evidence-based practice is the integration of the best available research with clinical expertise in the context of patient characteristics, culture, and preferences

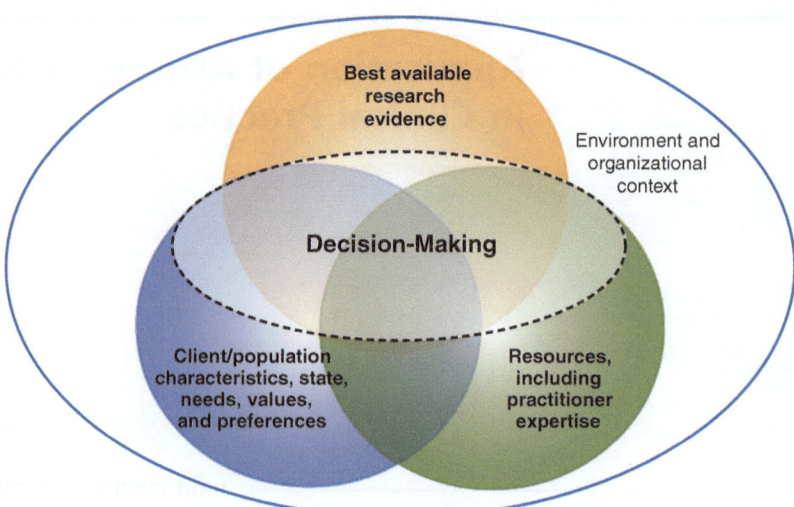

If there is not yet any disease-modifying therapy, what is the importance of an early diagnosis in routine clinical practice today? There are several reasons to make an early diagnosis even if we cannot modify the course of the disease. For instance, once people become demented, they can no longer plan for their future or dictate their end-of-life care. An early diagnosis of AD gives a person the opportunity to decide on important questions before he or she becomes demented (Martínez-Rivera et al. 2008; van Rossum et al. 2012). It also has important consequences for the patient's family. However, an early diagnosis of AD may also have negative psychological consequences in an otherwise well-functioning person who must now consider an inexorable decline towards a state of illness and dependency. Consequently, the pros and cons of early diagnosis must be carefully weighed for each individual prior to performing a confirmatory test.

Even if we have decided to use biomarkers to support the diagnosis of AD, there are many questions to be faced. It is important to emphasize that standardization of these biomarkers is currently limited, and results often vary from laboratory to laboratory. Ultimately, it will be necessary to interpret biomarker data in the context of well-established normative values. Moreover, procedures for acquisition and analysis of samples must be established to

implement these biomarker criteria on a broad scale. Although we consider biomarkers as "negative" or "positive" for purposes of classification, it is recognized that varying severities of an abnormality may confer different likelihoods or prognoses, which are difficult to quantify accurately for broad application. Currently it is hard to understand the relative importance of different biomarkers when used together and to interpret results when biomarker data conflict with one another.

Equally important, there are few truly predictive studies at the individual subject level or in unselected populations. The use of biomarkers in clinical practice will require the ability to assign a likelihood of progression for an individual person over a specific time interval through the use of a single or multiple biomarkers. Knowledge about the timing of decline is also severely limited because the ability to detect change depends on the period of observation or prediction. A complete understanding of the role of biomarkers in predicting decline will require both short-term and long-term periods of observation (Menéndez-González 2014a).

Finally, little is known about outcome when biomarkers provide conflicting results. When a panel of biomarkers is used, it is possible that for some individuals, one biomarker will be positive, one will be negative, and one equivocal.

The long-term significance of such findings may also vary with the length of follow-up.

Therefore, many questions have important repercussions for the management of patients suspected of suffering from AD:

- What biomarker is better for making an early diagnosis?
- Which biomarker is better for follow-up?
- Which biomarker is better for differential diagnosis?
- How should we interpret the results of these tests in coordination with clinical or genetic findings?
- How should we combine the results from different biomarkers?

The answers to these questions are not easy and many remain unanswered, relying on upcoming science. The rest of this chapter deals with the main challenges faced in the implementation of biomarkers in everyday clinical practice.

4.2 CSF Biomarkers

As seen in previous chapters, augmented CSF concentrations of p-Tau and reduced levels of amyloid-β (Aβ)1-42 have been replicated in a large number of studies with different clinical scenarios:

1. A decrease in CSF Aβ1-42 to about 50 % of the level in cognitively normal elderly subjects has been regularly reported, whereas an increase in CSF t-Tau to approximately 300 % of the level in cognitively normal elderly subjects and a less evident growth in CSF p-Tau to about 200 % have been repeatedly detected. Such biomarkers show 80–95 % sensitivity and specificity in the dementia phase of the pathology.

2. Prodromal AD CSF biomarkers also exhibit a high predictive value in detecting prodromal AD in MCI subjects. The increased ratio of Tau/Aβ1-42 and p-Tau/Aβ1-42 in normal subjects has been related to an amplified risk of conversion to MCI. A study with a protracted clinical follow-up period has revealed that the combination of all three core CSF biomarkers shows a sensitivity of 95 % in recognizing prodromal AD in MCI (van Harten et al. 2013).

3. Using a ratio of either p-Tau/Aβ1-42 or t-Tau/Aβ1-42 in differentiating AD from other dementias, the sensitivity was reported to be up to 92 % and specificity 86 %, with an overall accuracy of 90 % for the presence of pathologic neuritic plaque in the brain. Accuracy was particularly high using this combination of CSF biomarkers in differentiating AD from frontotemporal lobe dementia (FTLD), progressive supranuclear palsy (PSP), Parkinson's disease dementia (PDD), and cortical basal degeneration (CBD). It was less clear for dementia with Lewy bodies (DLB) and vascular dementia, possibly because of the propensity for mixed pathology in DLB and vascular dementia. Patients with Creutzfeldt-Jakob disease demonstrated extremely high CSF t-Tau with relatively normal levels of p-Tau and Aβ1-42.

However, several methodological limitations remain before biomarkers can be applied in clinical practice. The measurements of CSF concentrations of these biomarkers using enzyme-linked immunosorbent assays (for example, Innogenetics, Ghent, Belgium) or multiplex techniques (for example, xMAP, Luminex) have an acceptably low coefficient of intralaboratory variability (5–10 %), but the high interlaboratory variation (20–30 %) hinders comparison of data generated in different settings. Potential sources of variation include preanalytical conditions (that is, sample handling and aliquot storage), analytical conditions (different methods), and postanalytical norms for patients in defining cutoff points (that is, age or apolipoprotein 4 status). Furthermore, the current body of knowledge regarding biomarkers fails to categorize clinical scenarios characterized by ambiguous, indeterminate, or conflicting results involving multiple biomarkers.

As a result, globally recognized reference and cutoff values have not yet been established. Because global biomarker cutoff levels cannot be defined owing to the high extent of variability, each laboratory should employ internally validated cutoff values and guarantee longitudinal stability in its measurements. Progress in standardization of laboratory protocols, the enhancement of kit performance, and the use of fully automated tools are expected to improve the effectiveness of CSF AD biomarkers.

Another question to be taken into account regards the lumbar puncture (LP) procedure needed to take the CSF sample. There is a wide range of attitudes and beliefs about the convenience and feasibility of LP and its practical value in the management of patients. LP may be regarded as invasive or complicated and time-consuming, and patients may fear the procedure. One of the most controversial issues when discussing CSF biomarkers for early AD diagnosis has to do with the collection procedure itself. A debate exists about whether this technique can and should be used regularly, or if it is too risky for routine practice. Clinically, LP is performed routinely in clinics for various laboratory analyses to diagnose diseases such as meningitis, encephalitis, or inflammatory diseases like multiple sclerosis, as well as to inject spinal anesthetics or chemotherapy drugs. Nevertheless, many still feel that the benefits of its use for testing AD biomarkers do not outweigh the risks.

As a result, the use of LP for testing CSF biomarkers in the diagnosis of AD is surprisingly culturally dependent and subject to changes in fashion. Some clinicians support its use in daily clinical practice (Lanari and Parnetti 2009; Ariza-Zafra and Torrente-Orihuela 2005; Galuzzi et al. 2013; Menéndez-González 2014b) and in some countries LP is almost routine (e.g., Scandinavian countries, The Netherlands). In other areas (e.g., North America), it is regarded as very serious, to be used for research purposes only under strict protocols (Cummings 2011; Wilner 2010).

Some studies have already assessed the risks of LP, and the procedure seems to be both "safe and acceptable." In a multisite study in the United States, 342 people underwent 428 LPs. Adverse effects such as pain, anxiety, and the well-known post–lumbar puncture headache (PLPHA) were quantified and compared with controls. Overall, pain and anxiety levels were low as rated on a visual analog scale but generally were rated higher by younger normal subjects than by older participants. This theme remains true in studies looking at PLPHA frequency and severity: those who are younger (especially women) are at higher risk (Evans et al. 2000). PLPHA was unrelated to factors such as the position during the procedure (seated vs. lying), and the frequency of these headaches was lower in the MCI/AD group (age >60 year) than in any other subject group. This is a promising conclusion as far as AD is concerned, as all of the participants are older and many have MCI or AD. Other studies designed to assess LP procedures specifically in patients with AD demonstrated that LP performed with a 24G Sprotte atraumatic needle (blunt, "bullet" tip) was well tolerated, with good acceptability (Peskind et al. 2009).

As with many other medical techniques, the more often a procedure is done, the safer it becomes. To obtain more in-depth knowledge on the factors affecting the complications of LP for testing biomarkers in patients with cognitive impairment, the Alzheimer's Association is supporting a multicenter feasibility study to establish the incidence of PLPHA and other complications in individuals with cognitive disturbances and to learn the factors related to the occurrence of PLPHA, including the type of center, the experience of the physician, patient characteristics (e.g., diagnosis, cognitive function), the patient's attitude and knowledge about LP, and the LP procedure itself.

In summary, though LP seems safe and CSF biomarkers have made significant advances and hold great promise for future application in the clinical setting, recommendations for routine clinical practice cannot be reached for their general use until the limitations discussed above have been fixed. Working groups are in progress to develop protocols to resolve these issues. Indeed, the Alzheimer's Biomarkers Standardization Initiative has already published recommendations for CSF biomarker testing in AD (Table 4.1) (Vanderstichele et al. 2012).

Table 4.1 Alzheimer's biomarkers standardization initiative recommendations

Subject	Recommendation
CT scan or MRI performed before LP	LP should not be performed in patients with high intracranial pressure or a mass lesion in the brain
Concomitant medication	LP should not be performed in patients treated with anticoagulants (e.g., warfarin). Treatment with platelet aggregation inhibitors is not a contraindication
Diurnal variation	No diurnal variation
CSF gradient/volume	No gradient observed. No requirement for a certain fraction. Minimum volume of 1.5 mL
Meal consumption	No need for fasting
Position	LP may be performed with the patient either sitting or lying down. The position of the patient does not affect the results
Location	Vertebral body L3–L5. The incision point of the needle (L3–L4 or L4–L5) does not affect the results
Disinfection/anesthesia	Disinfection will reduce the risk of local infection. Local anesthetics introduce a risk of adverse effects, but can be given to patients who worry about local pain during LP
Needle	Use a small-diameter (0.7 mm and 22 G), preferably nontraumatic needle. A small-gauge needle will make a smaller hole in the dura, aiding healing, and an atraumatic needle will reduce the chance of blood contamination in the CSF
Rest	Leave the patient to rest for half an hour after LP. Prolonged bed rest or other procedures will not influence the risk of post-LP headache
Tubes and aliquotation (type, volume, homogeneity)	Each laboratory should use the same polypropylene tube. Glass or polystyrene tubes should never be used. Tubes of the smallest volume should be used, and these should be filled to at least 50 % of their volume
Documentation of sampling/aliquotation	It is important to have carefully recorded and validated details concerning each stored sample so that any investigator using the sample has its precise history
Centrifugation (speed and temperature)	Centrifugation is required only for visually hemorrhagic samples. Centrifuge as soon as possible, within 2-h of LP (on site or at nearest laboratory). Speed has no effect; recommend 2,000 g for 10 min at room temperature (controlled)
Time and temperature before storage	Samples may be sent by regular post (transport $\leq$5 days)
Method of freezing (liquid nitrogen, dry ice, slow freezing at −20 or −80 °C)	Freezing at −80 °C for storage. No difference between methods of freezing
Length of storage (when frozen)	Storage at −20 °C for <2 months. Note: no evidence of any effect for $\leq$2 years at −80 °C
Number of freeze/thaw cycles	Limit the number of cycles to 1 or 2
Interfering substances (hemolysis)	Traumatic LP: Discard first 1 to 2 mL. Samples with an erythrocyte count of 500/μL should not be used without centrifugation

Summary of recommendations for preanalytical and analytical aspects for Alzheimer's disease biomarker testing in cerebrospinal fluid (CSF)

LP lumbar puncture

(Adapted from Vanderstichele et al. 2012)

4.3 Neuroimaging Biomarkers

4.3.1 MRI

The advantage of imaging biomarkers versus CSF biomarkers can be found in their provision of information not only about the presence of a certain pathology but also about its topography and its actual extent in the brain. This advantage may be an important for disease staging, follow-up and therapy control, and differential diagnosis.

The quantification of medial temporal lobe atrophy (MTA) can be assessed using several different neuroimaging measurements, including rating scales, two-dimensional (2D) methods, and three-dimensional (3D)–volumetric methods.

Visual assessment rating scales are quick and can be performed on large numbers of scans in a clinical setting, but they lack objectivity.

Volumetric analysis provides an accurate and detailed measure of a predetermined circumscribed area or region of interest. For AD, the most used structure is the whole hippocampus. Manual volumetry is considered the gold standard, but it has some disadvantages. First, it requires training, as the tracer must learn to delineate the hippocampus's boundaries and the anterior and posterior limits. Then segmentation of the hippocampus takes approximately 20–30 min, depending on user experience, which limits routine clinical use. Some groups have automated segmentation techniques and protocols for multi-atlas–driven automatic segmentation of the hippocampus.

Two-dimensional methods such as the MTA index (MTAi) are very interesting for daily clinical practice because of their objectivity and quick and easy implementation, but they require further validation.

Novel imaging instrumentation such as hybrid PET/MR scanners may offer an additional opportunity to merge the complementary information from different imaging modalities into new, integrated, in vivo biomarkers of neurodegeneration. Suitable magnetic resonance procedures such as resting-state functional MRI (fMRI) or arterial spin labeling potentially may provide information on neuronal dysfunction relatively similar to FDG-PET findings, but the individual clinical value of these methods has not yet been established.

4.3.2 FDG PET

It is known that FDG mirrors neuronal dysfunction, and from a pathophysiological point of view, it appears obvious that functional changes should be detectable ahead of neuronal loss and brain atrophy. In fact, studies have been able to demonstrate that FDG PET has higher sensitivity than structural MRI in the prediction of AD at the stage of MCI. FDG PET may also be able to monitor changes in neuronal function in response to therapy, which may not be detectable with CSF measurements.

A plethora of data underline the clinical value of FDG PET for early and differential diagnosis of neurodegenerative disorders, as well as its complementary features when compared when amyloid imaging, so it can be expected that FDG PET will remain an important biomarker in the coming years.

4.3.3 Molecular Neuroimaging

Without doubt, amyloid imaging represents one of the most important biomarkers for scientific purposes. As seen in previous chapters, PET is a robust marker of neocortical fibrillar amyloid deposition in brain (Fig. 4.2) (Nordberg et al. 2013). However, there is wide individual variation in the brain amyloid load in MCI, and in the course of amyloid accumulation in relation to the clinical diagnosis of AD (Kemppainen et al. 2014).

Current scientific evidence suggests a useful application of amyloid imaging in patients with MCI, in AD with atypical presentation (e.g., early onset), and when the diagnosis is uncertain after evaluation by a dementia expert. MCI patients with positive amyloid-PET scans are at high risk for prodromal AD, whereas amyloid-PET

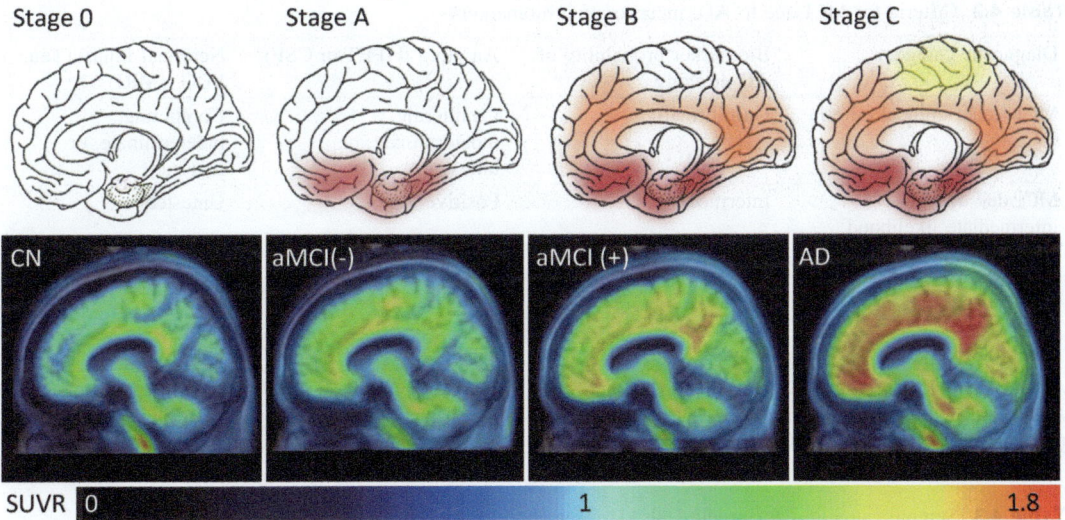

Fig. 4.2 Representative positron emission tomography (PET) scans of healthy controls, patients with mild cognitive impairment (MCI), and patients with Alzheimer's disease. The color scale reflects retention of [^{11}C]Pittsburgh Compound B (PiB): *red* high; *green* intermediate; *blue* low (from Nordberg et al. 2013; with permission)

negativity has around 100 % negative predictive value for progression to AD (Nordberg et al. 2013).

The establishment of this sophisticated method for in vivo assessment and quantification of a molecular neuropathology will certainly also depend on reimbursement issues and on whether trials of anti-amyloid therapy will be followed further and will yield promising pharmacotherapeutic approaches.

However, preclinical amyloid imaging also raises important ethical issues. Amyloid imaging is definitely useful to understand the role of Aβ in early stages, to define at-risk populations for research or clinical trials, and to assess the effects of anti-amyloid treatments, we are not yet ready to translate research results into clinical practice and policy.

4.4 Diagnostic Criteria and Impact on Clinical Outcomes

The latest proposed criteria for prodromal AD, MCI due to AD, and probable AD dementia incorporate molecular evidence of Aβ pathology and also consider measures of brain structure

and function as supportive biomarkers in research and clinical practice (Tables 4.2 and 4.3) (Albert et al. 2011; Sperling et al. 2011).

According to these guidelines, biomarkers are suited to play an important diagnostic role in all stages of disease. In short, the proof of amyloid pathology (as possible with amyloid PET or CSF Aβ42), accompanied by proof of neuronal injury (as hippocampal atrophy or CSF Tau) or cerebral hypometabolism (as possible with FDG PET) and finally proof of cognitive impairment, sum up to an increasing probability for AD. After prepathology stage characterized by normal biomarkers and absence of cognitive impairment, AD dimensionally (not categorically) emerges through an asymptomatic stage (biomarkers abnormal, no cognitive impairment) and then a symptomatic stage (biomarkers abnormal, cognitive impairment), which can be further differentiated into a subjective cognitive impairment (SCI), MCI, and finally a syndromal dementia stage (AD dementia). Notably, these categories are merely restrictive research or practical clinical constructs and should not mask the true, continuous dimensional character of AD. Individuals with preclinical AD can be categorized into three stages using cognitive

Table 4.2 Criteria for MCI due to AD, incorporating biomarkers

Diagnostic category	Biomarker probability of AD etiology	Amyloid β (PET or CSF)	Neuronal injury (Tau, FDG, sMRI)
MCI—core clinical criteria	Uninformative	Conflicting, indeterminate, or untested	Conflicting, indeterminate, or untested
MCI due to AD—intermediate likelihood	Intermediate	Positive	Untested
		Untested	Positive
MCI due to AD—high likelihood	Highest	Positive	Positive
MCI—unlikely due to AD	Lowest	Negative	Negative

AD Alzheimer's disease; *CSF* cerebrospinal fluid; *FDG* fluorodeoxyglucose; *MCI* mild cognitive impairment; *PET* positron emission tomography; *sMRI* structural MRI
(Adapted from Albert et al. 2011)

Table 4.3 Staging categories for preclinical AD research

Stage	Description	Amyloid β (PET or CSF)	Markers of neuronal injury (Tau, FDG, sMRI)	Evidence of subtle cognitive change
Stage 1	Asymptomatic cerebral amyloidosis	Positive	Negative	Negative
Stage 2	Asymptomatic amyloidosis + "downstream" neurodegeneration	Positive	Positive	Negative
Stage 3	Amyloidosis + neuronal injury + subtle cognitive/behavioral decline	Positive	Positive	Positive

AD Alzheimer's disease; *CSF* cerebrospinal fluid; *FDG* fluorodeoxyglucose (18F); *MCI* mild cognitive impairment; *PET* positron emission tomography; *sMRI* structural MRI
(Adapted from Sperling et al. 2011)

markers and biomarkers: Those showing only anomalous amyloid markers are classified in stage 1; those with both atypical amyloid and injury markers are considered to be in stage 2; and those showing both unusual amyloid and injury markers accompanied by minimal cognitive impairments, such as SCI, are classified in stage 3 (Fig. 4.3) (Lazarczyk et al. 2012). According to this model, the group currently defined as "preclinical AD" is heterogeneous and comprises two subpopulations. Firstly, there is the group of individuals at different stages of preclinical AD defined by the biomarkers indicated in the lower panel of Fig. 4.3. All of these individuals will progress to dementia, and we call this phase "presymptomatic AD." The second group comprises individuals who are positive for amyloid markers and neuronal injury markers, and fall into one of the stages of preclinical AD, based on the current classification. However, this population has efficient active compensatory mechanisms, and remains resistant to dementia (stable asymptomatic cerebral amyloidosis).

Some studies have assessed the impact of using biomarkers in clinical practice, as well as the concordance between results from different biomarkers. For instance, the concordance between FDG and amyloid scan results is high. However, PET had a moderate effect on clinical outcomes. Discordant PiB PET scans have a greater effect than discordant FDG PET scans, and their influence on diagnosis is greater than on treatment (Sánchez-Juan et al. 2013).

Further prospective studies are needed to better characterize the clinical impact of using biomarkers in the management of patients with cognitive impairment.

Fig. 4.3 Hypothetical model of preclinical Alzheimer's disease (AD). *sMRI* structural MRI (from Lazarczyk et al. 2012)

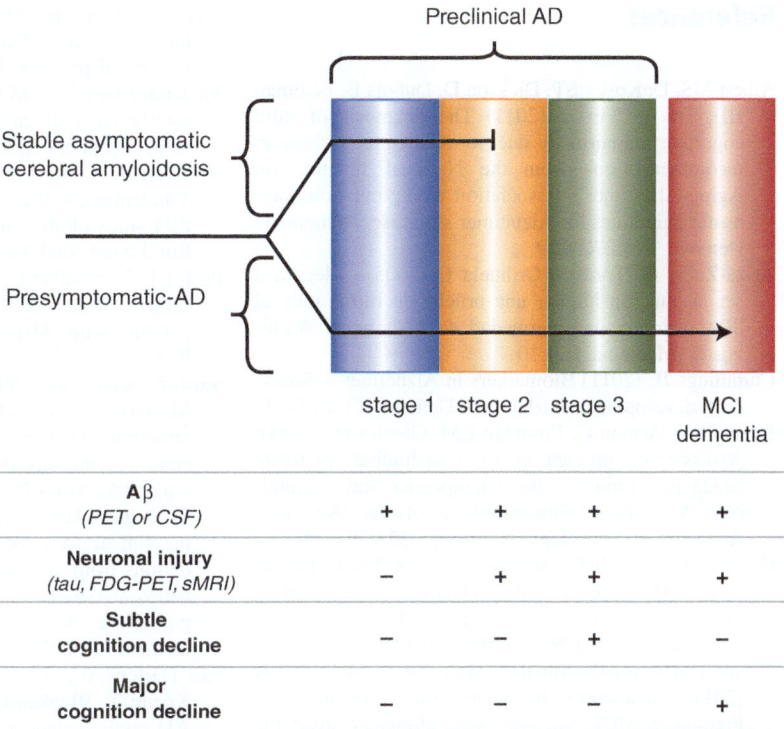

	stage 1	stage 2	stage 3	MCI dementia
Aβ *(PET or CSF)*	+	+	+	+
Neuronal injury *(tau, FDG-PET, sMRI)*	−	+	+	+
Subtle cognition decline	−	−	+	−
Major cognition decline	−	−	−	+

4.5 Key Concepts

1. Although lumbar puncture is safe and CSF biomarkers provide good predictive values and have been replicated by a large number of studies in different clinical scenarios, several methodological limitations should be solved before biomarkers can be recommended in clinical practice.
2. Neuroimaging techniques are accessible and are already being used in some centers with different approaches:
   - Visual assessment rating scales are quick but lack objectivity.
   - Volumetric analysis provides an accurate and detailed measure but is time-consuming and requires extensive training.
   - 2D methods are interesting for routine clinical practice because of their objectivity and quick and easy implementation, but they need further validation.
3. FDG-PET may be used in clinical practice for early and differential diagnosis of neurodegenerative disorders and may also be used to monitor progression.
4. A plethora of data underline the clinical value of FDG PET for early and differential diagnosis of neurodegenerative disorders.
5. There is wide individual variation in the brain amyloid load in patients with MCI.
6. Amyloid-positive scans in MCI patients are an indicator of high risk for prodromal AD, whereas negative amyloid scans have a 100 % negative predictive value for progression to AD.
7. Amyloid imaging may be useful to define at-risk populations for research or for clinical trial, and to assess the effects of anti-amyloid treatments, but its role in clinical practice is not so clear.
8. Individuals with preclinical AD can be categorized into three stages, using cognitive markers and biomarkers.

References

Albert MS, DeKosky ST, Dickson D, Dubois B, Feldman HH, Fox NC et al (2011) The diagnosis of mild cognitive impairment due to Alzheimer's disease: recommendations from the National Institute on Aging-Alzheimer's Association workgroups on diagnostic guidelines for Alzheimer's disease. Alzheimers Dement. 7(3):270–279

Ariza-Zafra G, Torrente-Orihuela C (2005) ¿Llegará a ser la punción lumbar una prueba de rutina para el diagnóstico de la enfermedad de Alzheimer? Archivos de Medicina 1(2):10

Cummings JL (2011) Biomarkers in Alzheimer's disease drug development. Alzheimers Dement 7(3):e13–e44

Evans RW, Armon C, Frohman EM, Goodin DS (2000) Assessment: prevention of post-lumbar puncture headaches: report of the Therapeutics and Technology Assessment Subcommittee of the American Academy of Neurology. Neurology 55:909–914

Galluzzi S, Geroldi C, Amicucci G, Bocchio-Chiavetto L, Bonetti M, Bonvicini C et al (2013) Supporting evidence for using biomarkers in the diagnosis of MCI due to AD. J Neurol 260:640–650

Heister D, Brewer JB, Magda S, Blennow K, McEvoy LK (2011) Alzheimer's disease neuroimaging initiative. Predicting MCI outcome with clinically available MRI and CSF biomarkers. Neurology 77:1619–1628

Kemppainen NM, Scheinin NM, Koivunen J, Johansson J, Toivonen JT, Någren K et al (2014) Five-year follow-up of (11)C-PIB uptake in Alzheimer's disease and MCI. Eur J Nucl Med Mol Imaging 41:283–289

Lanari A, Parnetti L (2009) Cerebrospinal fluid biomarkers and prediction of conversion in patients with mild cognitive impairment: 4-year follow-up in a routine clinical setting. Sci World J 9:961–966

Lazarczyk MJ, Hof PR, Bouras C, Giannakopoulos P (2012) Preclinical Alzheimer disease: identification of cases at risk among cognitively intact older individuals. BMC Med 10:127

Martínez-Rivera M, Menéndez-González M, Pérez-Piñera P (2008) Biomarcadores para la Enfermedad de Alzheimer y otras demencias degenerativas. Archivos de Medicina 4(3):3

Menéndez-González M (2014a) The many questions on the use of biomarkers for neurodegenerative diseases in clinical practice. Frontiers Aging Neurosci 8:45

Menéndez-González M (2014b) Routine lumbar puncture for the early diagnosis of Alzheimer's disease. Is it safe?. Frontiers Aging Neurosci 8:65

Nordberg A, Carter SF, Rinne J, Drzezga A, Brooks DJ, Vandenberghe R et al (2013) A European multicentre PET study of fibrillar amyloid in Alzheimer's disease. Eur J Nucl Med Mol Imaging 40:104–114

Peskind E, Nordberg A, Darreh-Shori T, Soininen H (2009) Safety of lumbar puncture procedures in patients with Alzheimer's disease. Curr Alzheimer Res 6:290–292

Sánchez-Juan P, Ghosh PM, Hagen J, Gesierich B, Henry M, Grinberg LT et al (2013) Practical utility of amyloid and FDG-PET in an academic dementia center. Neurology 18 Dec 2013 [Epub ahead of print]

Sperling RA, Aisen PS, Beckett LA, Bennett DA, Craft S, Fagan AM et al (2011) Toward defining the preclinical stages of Alzheimer's disease: recommendations from the National Institute on Aging-Alzheimer's Association workgroups on diagnostic guidelines for Alzheimer's disease. Alzheimers Dement 7(3):280–292

van Harten AC, Smits LL, Teunissen CE, Visser PJ, Koene T, Blankenstein MA et al (2013) Preclinical AD predicts decline in memory and executive functions in subjective complaints. Neurology 81:1409–1416

van Rossum IA, Vos SJ, Burns L, Knol DL, Scheltens P, Soininen H et al (2012) Injury markers predict time to dementia in subjects with MCI and amyloid pathology. Neurology 79:1809–1816

Vanderstichele H, Bibl M, Engelborghs S, Le Bastard N, Lewczuk P, Molinuevo JL et al (2012) Standardization of preanalytical aspects of cerebrospinal fluid biomarker testing for Alzheimer's disease diagnosis: a consensus paper from the Alzheimer's biomarkers standardization initiative. Alzheimers Dement 8:65–73

Wilner AN (2010) Alzheimer's CSF test: useful or useless? Medscape. http://www.medscape.com/viewarticle/730235_3. Accessed 17 Jan 2014

Index

M. Menéndez González, *Atlas of Biomarkers for Alzheimer's Disease*,
DOI: 10.1007/978-3-319-07989-9, © Springer International Publishing Switzerland 2014

GPSR Compliance

*The European Union's (EU) General Product Safety Regulation (GPSR)
is a set of rules that requires consumer products to be safe and our
obligations to ensure this.*

*If you have any concerns about our products, you can contact us on
ProductSafety@springernature.com*

In case Publisher is established outside the EU, the EU authorized
representative is:

Springer Nature Customer Service Center GmbH
Europaplatz 3
69115 Heidelberg, Germany

Batch number: 09467656

Printed by Printforce, the Netherlands